Pleurotus nebrodensis
A Very Special Mushroom

Edited By

Maria Letizia Gargano

Department of Agricultural and Forest Sciences
Università di Palermo
viale delle Scienze 11
I-90128 Palermo
Italy

Georgios I. Zervakis

Agricultural University of Athens
Laboratory of General and Agricultural Microbiology
Iera Odos 75
11855 Athens
Greece

&

Giuseppe Venturella

Department of Agricultural and Forest Sciences
Università di Palermo
viale delle Scienze 11
I-90128 Palermo
Italy

CONTENS

FOREWORD

It is a great pleasure for the Species Survival Commission of IUCN (the International Union for Conservation of Nature) to introduce this noteworthy ebook about *Pleurotus nebrodensis*, edited by M.-L. Gargano, G. I. Zervakis and G. Venturella for several reasons.

Firstly, the Fungi Kingdom was, until now, largely underrepresented in the conservation world. Fortunately, this situation seems to have improved gradually, as shown, for instance, by the recent creation of four fungi specialist groups within IUCN SSC (five if we also consider the Lichens specialist group). As a first result, the knowledge on the conservation status of several mushrooms is now better known in proportion to their assessments with the IUCN Red List criteria. There are of course still many things to do in order to better insure their conservation, but the mycologists' voice has begun to be heard.

Secondly, because this ebook is an exceptional blend of different approaches focused on one species conservation. Many examples show that conservation measures for a species are successful only if all factors influencing its habitat and life are studied and taken into account, including of course the social and economics ones, which are unfortunately too often neglected. And this has been particularly well understood by Prof. Venturella and his team, who have not only tirelessly studied this species and its environment - *sensu lato* - since many years, but have also planned very detailed cultivation methods in order to alleviate the pressure on *Pleurotus nebrodensis* by its over collection in the wild. IUCN's long experience in conservation throughout the world demonstrates that the most successful stories are the ones that benefit the local communities. And this is precisely the case in the *Pleurotus* conservation strategy.

This makes us very confident in the long term sustainable management of this mushroom.

Finally, because this story owes its origin to in Sicily, a region deeply attached to its cultural and natural heritage, for thousands of years, nature in Sicily has been shaped by people who have known how to use sustainably its natural resources.

Even if this is less the case today, due to the intensive agriculture, the infrastructures, the urbanization and the development of mass tourism, Sicilians are still close to their nature and ready to conserve and value it.

We hope that, in the near future, the threat category of *Pleurotus nebrodensis*, which is currently CR (Critically Endangered), will decrease due to the conservation measures implemented and expect that this success story will be an example for other threatened fungi and plants.

Neuchâtel

Bertrand de Montmollin
IUCN - Species Survival Commission
Switzerland

PREFACE

A progressive decrease in the levels of biodiversity within ecosystems characterizes the ongoing scenario for our planet.

Politicians, academics, networks, media and organizations dealing with the conservation of nature primarily directed their attention to cases of threatened animals and plants at risk of extinction.

All these organisms have a strong impact on public opinion because of their physical size, notable presence and/or distribution.

It is therefore more difficult to raise the level of public attention to seemingly "lower" organisms such as fungi.

The problem that I encountered about twenty years ago when I started to study *Pleurotus nebrodensis* was: How a mushroom unknown to the vast majority of people could be put under the spotlight of public opinion?

Back in 1863, Giuseppe Inzenga had experienced a similar problem when he began to observe and describe a white *Pleurotus* mushroom, since he could not explain how "a species so easily distinguishable from the others would have missed the eye of skilled botanists of that time".

Inzenga understood that the fungus could be a different species from those already described in literature and this was confirmed when Elias Fries provided his authoritative opinion.

After Inzenga's death (1887) and up until 1995, no researcher could effectively deal with the species described by Inzenga under the binomial *Agaricus nebrodensis* Nobis [current name *Pleurotus nebrodensis* (Inzenga) Quél.]. In the meantime, the nomenclatural status of this taxon has undergone various changes ranging between synonyms and placements in the rank of "variety" or "subspecies" of *Pleurotus eryngii* (DC.) Quél.

The efforts made in recent years by the undersigned and his collaborators to gather information useful for the preparation of a dossier to be submitted to IUCN (International Union for Conservation of Nature) have delivered in the year 2005 a prestigious goal, *i.e.* the inclusion of *P. nebrodensis* in the IUCN Red List of Threatened Species and the Top 50 Mediterranean Island Plants. The scientific community has spent much time before confirming that *P. nebrodensis* is a species at risk of extinction, while since 1600, the Sicilian population enjoyed the excellent organoleptic qualities of these mushrooms which gained a high commercial value.

Known locally as "fungo di basilisco", derived from vernacular name of the plant on the roots of which the fungus is fruiting, *P. nebrodensis* is still today an important source of income particularly for the community of Madonie Mts (Northern Sicily).

Unfortunately, the overexploitation and the consequent excessive human pressure on the habitats of growth, stimulated by the high sale price of around 50 euro per kilo, led to a high risk of extinction for the fungus.

There are still some elements missing, in the long process of assembling the pieces of the jigsaw puzzle for the final construction of a "protection shield" for *P. nebrodensis*, such as a greater awareness of the value of this natural resource and more effective *in situ* and *ex situ* conservation actions.

Besides, there is also a need to improve the process of mushroom cultivation in order to allow the market to lower prices which would subsequently alleviate part of the human pressure exerted on natural habitats of *P. nebrodensis* growth.

This e-book is yet another opportunity to disseminate the history and the quality of *Pleurotus nebrodensis* to a wide range of readers in order to ensure that the original vision of Giuseppe Inzenga would result in an opportunity for future generations to adopt this beautiful mushroom and conserve its wild populations, while rural communities would benefit by obtaining an additional income from a natural resource in a sustainable manner.

Giuseppe Venturella
Department of Agricultural and Forest Science
Università di Palermo
viale delle Scienze 11
I-90128 Palermo
Italy

List of Contributors

Maria Letizia Gargano

Agricultural and Forest Sciences, Università di Palermo, viale delle Scienze 11, I-90128 Palermo, Italy

David Minter

International Society for Fungal Conservation, 4 Esk Terrace, Whitby, North Yorkshire, YO21 1PA, UK

Elias Polemis

Agricultural University of Athens, Laboratory of General and Agricultural Microbiology, Iera Odos 75, 11855 Athens, Greece

Alessandro Saitta

Agricultural and Forest Sciences, Università di Palermo, viale delle Scienze 11, I-90128 Palermo, Italy

Giuseppe Venturella

Agricultural and Forest Sciences, Università di Palermo, viale delle Scienze 11, I-90128 Palermo, Italy

Georgios I. Zervakis

Agricultural University of Athens, Laboratory of General and Agricultural Microbiology, Iera Odos 75, 11855 Athens, Greece

CHAPTER 1

Fungal Conservation and Sustainability in Europe

David Minter[*]

International Society for Fungal Conservation, 4 Esk Terrace, Whitby, North Yorkshire, YO21 1PA, UK

Abstract: The Convention on Biological Diversity requires every participating country to prepare a national biodiversity strategy and action plan and submit periodic progress reports on efforts to achieve the plan's goal. The most recent such documents submitted by each participating country in Europe were evaluated from the perspective of fungal conservation using simple criteria. In every case, coverage of fungi was markedly poorer than for animals and plants, being mostly far less than 5% that of animals alone. No report dealt adequately with fungi. Most overlooked fungal habitats and contained no strategic planning explicitly for protecting fungi. Almost all confused fungi with plants. Some totally failed to mention fungi. At a governmental level, European fungal conservation and sustainability is deeply deficient. Ways to ameliorate this inadequate treatment through science, infrastructure, education and political approaches are discussed.

Keywords: *Ascomycota*, *Basidiomycota*, Biodiversity, Checklists, Convention on Biological Diversity, Education, Europe, Fungal conservation, Fungal habitat, Fungi, Important Fungal Areas, *in situ* Conservation, IUCN, Legal protection, Mycological literature, On-line digital products, *Pleurotus nebrodensis*; Redlists, Sustainability, Taxonomy.

INTRODUCTION

Pleurotus nebrodensis, the subject of this ebook, has a unique place in conservation history.

It was the first mushroom recognized as critically endangered by the *International Union for Conservation of Nature* [IUCN]. Getting this fungus added to the IUCN's red list of endangered species was a great achievement and reflects the

---

*****Address correspondence to David Minter:** International Society for Fungal Conservation, 4 Esk Terrace, Whitby, North Yorkshire, YO21 1PA, UK; Tel: 0044(0)1491829031; E-mail: d.minter@cabi.org

Maria Letizia Gargano, Georgios I. Zervakis and Giuseppe Venturella (Eds)

important role Italian mycologists have played in fungal conservation. As an introduction to this ebook, I have been asked to place that achievement in the context of a wider review of fungal conservation and sustainability in Europe. My theme is at the same time large because there is so much to do, and small because so little has yet been done. Conservation is, uniquely, a mixture of science and politics.

The scientific component identifies threats to species and provides evidence of their decline, while the political component says, "something must be done about this". The present review cannot possibly cover all aspects of European fungal conservation, so I have chosen to focus my attention on political aspects and, in particular, what has been done, and what should be done to protect fungi at a governmental level. For the purposes of this chapter, Europe is defined as the countries of the continental landmass west of the Urals and north of the Caucasus watershed, and of its associated islands (Iceland, Ireland and the UK), including countries partly outside that area (Russia and Turkey), and countries outside that area but inside the European Union (Cyprus).

THE RIO CONVENTION

The *Convention on Biological Diversity* (CBD) makes a good starting point.

It resulted from the 1992 international summit in Rio de Janeiro, which acknowledged the importance of biological diversity and recognized that all living species have the right to live on this planet. The CBD came into force in 1993, and more than 190 countries are now signatories, including virtually all European countries (Andorra and the Holy See are Exceptions). Its main goal is the conservation of biological diversity, and the convention requires participating countries to produce national biodiversity strategies and action plans explaining how they propose to achieve that goal, together with periodic progress reports. There are hundreds of these documents. All of them are substantial in size, often well over 100 pages long. Most are written in English and are freely and openly available from the CBD website [www.cbd.int] as PDF or MS-WORD files. These action plans and reports provide a remarkable insight into what governments, in their own words, see as their duties in respect of biodiversity conservation.

Objectivity. Before looking at the documents themselves, it is helpful to form an independent view of what they should contain, so that there is some standard against which each can be objectively compared. The first point to make is that these documents deal with all aspects of biological diversity. The term "biological diversity" is defined in the text of the CBD as "the variability among living organisms from all sources including, inter alia, terrestrial, marine and other aquatic ecosystems and the ecological complexes of which they are part; this includes diversity within species, between species and of ecosystems". It can be seen that this text is neutral. It does not allocate a greater importance to one group of organisms over any other. The smallest species of worm has the same right to conservation as the largest tree. The action plans and reports should therefore provide a roughly equal and balanced consideration of all groups of organisms which occur in the country in question.

The individual documents cannot possibly deal with every detail of biological diversity. They are, after all, summaries. One might expect to see some illustrative examples, but not exhaustive lists of species with separate detailed protection plans for each. For many groups of organisms, new species are still being discovered, and for some groups only a few of the species which probably exist have yet been named. Furthermore some groups (for example flowering plants and vertebrates) are better known than others (for example algae, bacteria, fungi, invertebrates, non-flowering plants and protozoa), so the illustrative examples are more likely to be from better known groups. Similarly, there is usually more information about conservation actions for better known groups.

Where a group has been neglected in the past, such information is often lacking.

As a result, action plans and reports are likely to contain a lot of information about the conservation status of flowering plants and vertebrates, and about actions being taken to address known deficiencies in protecting them. But that doesn't mean it's acceptable to say nothing about the other groups.

These documents are supposed to address the conservation of all biological diversity, not just of the bits already known. We should therefore expect expensive coverage of lesser known groups.

For lesser known groups, they should certainly contain a full review of what is not known, with detailed plans of how to fill the knowledge gap, and a consideration of resources necessary to achieve that aim.

ASSESSING CBD DOCUMENTS

In this chapter, coverage of fungi is compared with coverage of animals and plants. To do that, the most recent documents submitted to the CBD by each of 44 European countries (including the most recent report from the European Union [EU] as a whole) were examined. In some cases, the document was a biodiversity strategy and action plan. In others it was a report. Using standard PDF-viewing software, with the options set to include both capital and lower case letters, each document was searched. In most cases, the search was for words, but sometimes it was more appropriate to search for particular combinations of letters. For example searching for "myc" found many different words relating to fungi, such as "mycology", "basidiomycete", "mycobiota" *etc*.

For coverage of animals, a search was made for the following words: "amphibian", "animal", "bird", "fauna", "fish", "insect", "mammal", "reptile" and "vertebrate" (which also found "invertebrate").

For plants, the search was for the following words or letter combinations: "bryo" (as in "bryophyte" but with animal words like "bryozoan" filtered out), "conifer", "fern", "flora", "flower", "moss" and "plant" (but excluding phrases like "sewage treatment plant").

For fungi, the search was for the following words or letter combinations: "fung" (as in "fungal", "fungi" and "fungus"), "lichen" (but excluding German words like "wissenschaftlichen"), "myc", "mould", "mushroom", "toadstool", "truffle" and "yeast". Where the document was in a language other than English (usually French, Russian or Spanish), the nearest translated equivalent was used, taking care with inflected languages to search only for the word stem.

The English language has rather few vernacular words for fungi and, in searching for fungal coverage, most were used. That was not the case for animals or plants, where many other vernacular names are available.

Taking the example of the animals, other suitable words which could also have been used include arthropod, bat, bear, bee, beetle, butterfly, cattle, coral, crustacean, dolphin, dragonfly, duck, falcon, frog, goose, lynx, mollusc, owl, porpoise, raptor, rodent, seal, snail, snake, tortoise, turtle, wader, wasp, whale, wolf and worm.

Even before the searches were made, therefore, it was clear that coverage of animals and plants in these documents would be under-estimated. For each search, the number of hits was recorded.

After the searches had been made, those numbers were bulked to produce overall totals for animals, fungi and plants from each country.

Table **1** summarizes the results.

Table 1: Number of times animal, fungal and plant words occurred in CBD country reports from Europe

Country [document file name; date received by CBD]	Animals	Fungi	Plants
Albania [al-nr-04-en.pdf; 1 Apr. 2011]	105	3	46
Austria [at-nr-04-en.pdf; 21 Oct. 2010]	104	26	85
Belarus [by-nbsap-v2-en.pdf; 6 Jan. 2011]	101	3	44
Belgium [be-nr-04-en.pdf; 5 Oct. 2009]	201	18	145
Bosnia and Herzegovina [ba-nbsap-01-en.pdf; 15 April 2011]	95	60	75
Bulgaria [bg-nr-04-en.pdf; 15 Oct. 2010]	193	23	221
Croatia [hr-nbsap-v2-en.pdf; 16 Jun. 2009]	249	9	56
Cyprus [cy-nr-04-en.pdf; 7 Sep. 2010]	130	5	94
Czech Republic [cz-nr-04-p1-en.pdf; 7 May 2009]	213	11	140
Denmark [dk-nr-04-en.pdf; 12 Feb. 2010]	309	21	144
Estonia [ee-nr-04-en.pdf; 5 Dec. 2008]	100	20	76
European Union [eur-nr-04-en.pdf; 15 May 2009]	296	2	58
Finland [fi-nr-04-en.pdf; 24 Jun. 2009]	222	90	245
France [fr-nbsap-v2-en.pdf; 20 May 2011]	22	2	16
FYROM [mk-nr-04-en.pdf; 26 Mar. 2010]	117	3	34
Germany [de-nr-04-en.pdf; 30 Apr. 2010]	140	5	153
Greece [gr-nr-03-en.pdf; 8 Apr. 2008]	119	1	138
Hungary [hu-nr-04-en.pdf; 8 Jun. 2009]	126	6	126

Table 1: contd....

Country [document file name; date received by CBD]	Animals	Fungi	Plants
Iceland [is-nr-02-en.pdf; 10 Mar. 2003]	14	0	14
Ireland [ie-nbsap-v2-en.pdf; 17 Jan. 2012]	126	2	26
Italy [it-nbsap-01-en.pdf; 22 Dec. 2010]	183	8	110
Latvia [lv-nr-04-en.pdf; 24 Mar. 2010]	103	11	96
Liechtenstein [li-nr-04-en.pdf; 10 Mar. 2010]	150	3	117
Lithuania [lt-nr-04-en.pdf; 23 Oct. 2009]	464	95	266
Luxembourg [lu-nr-04-fr.pdf; 9 Dec. 2009]	46	0	50
Malta [mt-nr-04-en.pdf; 2 Jun. 2010]	435	8	319
Moldova [md-nr-04-en.pdf; 16 Jun. 2009]	189	11	275
Monaco [mc-nr-04-fr.pdf; 30 Oct. 2009]	77	0	30
Montenegro [me-nr-04-en.pdf; 25 Oct. 2010]	243	22	166
Netherlands [nl-nr-04-en.pdf; 14 Apr. 2010]	80	6	53
Norway [no-nr-04-en.pdf; 24 Apr. 2009]	178	11	22
Poland [pl-nbsap-v2-en.pdf; 1 Jun. 2009	47	11	45
Portugal [pt-nr-04-en.pdf; 29 Sep. 2009]	400	4	319
Romania [ro-nr-04-en.pdf; 5 Oct. 2009]	166	5	79
Russia (including Russia in Asia) [ru-nr-04-ru.pdf; 9 Nov. 2009]	484	28	238
Serbia [rs-nbsap-01-en.pdf; 16 Mar. 2011]	257	33	186
Slovakia [sk-nr-04-en.pdf; 30 Sep. 2009]	183	26	188
Slovenia [si-nr-04-en.pdf; 7 Apr. 2011]	387	10	126
Spain [es-nbsap-v3-es.pdf; 30 Jan. 2012]	221	0	42
Sweden [se-nr-04-en.pdf; 8 Apr. 2009]	282	28	175
Switzerland [ch-nbsap-v2-en.pdf; 2 May 2012]	152	12	83
Turkey (including Turkey in Asia) [tr-nr-04-en.pdf; 28 Sep. 2010]	226	4	154
Ukraine [ua-nr-04-ru.pdf; 9 Jun. 2010]	141	21	171
UK [gb-nr-04-en.pdf; 20 May 2009]	292	94	347

This simple analysis revealed a huge imbalance. The EU report, for example, mention of animals was about 150 times greater than mention of fungi. In several cases there was no mention of fungi at all. For more than half of the countries considered, coverage of the fungi was at most 5% of that for animals. Only three countries exceeded 25%: the UK report (just over 30%), Finland (about 40%) and, with the highest coverage for fungi, Bosnia and Herzegovina, at just over 60% of that country's coverage of animals.

The imbalance would undoubtedly have been much greater if, as already noted, information about animals and plants had not been consistently under-estimated.

That analysis simply looked at the occurrence of particular words and combinations of letters. It did not look at the context in which they appeared.

To consider that context for the fungi, each document listed in Table **1** was examined for every mention of the fungal words and letter combinations used in the previous searches, and five simple questions were asked, each expecting a "Yes/No" answer.

- **Question 1**. Were fungi mentioned? **Comment**. Fungi may from time to time be inconvenient for humans, for example as parasites or agents of biodegradation, but that does not remove from them their right to live on this planet. If fungi were mentioned only as an exploitable resource, or as threats to other organisms, for example by reference to fungicides *etc*., the answer was judged to be negative.

- **Question 2**. Were fungi (including lichen-forming fungi) clearly, consistently and explicitly recognized as different from animals, plants and other biological kingdoms? **Comment**. It is a basic prerequisite of any report that the science is correct. Fungi have been generally recognized as belonging in their own independent biological kingdom since at least the late 1960s.

- **Question 3**. Was strategic consideration explicitly given to fungal conservation (example indicators: separate texts devoted to fungal conservation; lists of important fungus areas / fungal biodiversity hotspots; deficiencies in legal protection for fungi identified and plans present to rectify those deficiencies; threats to fungi identified; fungal red lists mentioned)? **Comment**. An assumption that fungi will somehow magically be conserved if other organisms are conserved is not good enough. There have to be explicit plans to deal with the particular issues raised in fungal conservation.

- **Question 4**. Were principal fungal habitats and roles taken into account (for example decomposers, dung fungi, endobionts, freshwater fungi, fungi on man-made products, fungi on naturally occurring inanimate substrata, lichen-forming fungi, marine fungi, mycorrhizal fungi, parasitic fungi)? **Comment**. Fungal habitats are different from animal or plant habitats, and need separate consideration.

- **Question 5**. Was the knowledge gap for fungi recognized with plans to address the problem? **Comment**. Pretending this problem does not exist is no solution. Both parts of the question need to be "yes" to get a positive score. Recognizing a knowledge gap and then failing to deal with it is not good enough.

For each question, a "yes" answer indicated that the report was adequate. A "yes" did not mean the report did well, only that it met the most basic requirement. To be providing adequate basic cover of fungal conservation, a "yes" answer to every question was necessary. Each answer of "no" indicated that the report was deficient for fungal conservation in an important aspect. No report scored five "yes" answers. Every report was deficient in at least one aspect. Most were deficient in three or more aspects. The results are shown in Appendix 1. To make evaluations rapidly understandable, an emoticon expressing disappointment is shown for each negative answer. The details are discussed below.

MENTIONING FUNGI AND GETTING THE TAXONOMY RIGHT

The reports of Iceland, Luxembourg, Monaco and Spain totally failed to mention fungi. The European Union, France and Greece mentioned fungi, but so fleetingly that there was no evidence of the document authors being aware of the status of fungi.

The documents produced by Austria, Belarus, Belgium, Bulgaria, Cyprus, the Czech Republic, Denmark, Italy, Latvia, Malta, Moldova, Montenegro, the Netherlands, Poland, Portugal, Romania, Slovakia, Slovenia, Turkey and Ukraine contained clear evidence that their compilers simply did not understand that fungi are a separate biological kingdom.

All to a greater or lesser extent confused fungi and particularly lichens with plants. Finland, Lithuania, Sweden and the UK clearly recognized that fungi are different, but included fungi in their contribution to the Global Strategy for Plant Conservation.

Only Bosnia and Herzegovina, Croatia, Estonia, Ireland, Russia, Serbia and Switzerland correctly treated fungi as different, with no confusion at all with other biological kingdoms.

STRATEGIC CONSIDERATION OF FUNGAL CONSERVATION

Governmental funding for fungal conservation was not mentioned in any of the documents examined. Where existing checklists and red lists were discussed, the way they were or would be funded was not described. The reports of Albania, Austria, Belarus, Belgium, Cyprus, Czech Republic, European Union, France, FYROM, Germany, Greece, Hungary, Iceland, Italy, Latvia, Luxembourg, Malta, Moldova, Monaco, Netherlands, Norway, Portugal, Romania, Russia, Slovakia, Spain, Turkey and Ukraine showed no evidence at all of any strategic consideration of fungal conservation. The reports of Croatia, Bulgaria, Denmark, Estonia, Ireland, Poland and Switzerland showed some minimal awareness that fungal conservation needs attention: Bulgaria, Croatia, Estonia, Ireland and Switzerland referred to fungal red lists, and Croatia, Denmark and Poland referred to threats to fungi, but in every case the mention was fleeting, without even the most superficial consideration of the issues inherent in each topic. Only Bosnia and Herzegovina, Finland, Lithuania, Serbia, Sweden and the UK made any effort to address conservation issues specific to fungi, and even in those cases coverage was very inadequate, and disproportionately small compared with that of animals and plants.

The report for Bosnia and Herzegovina contained separate paragraphs devoted to fungal conservation, and specifically about the generation of fungal checklists and red lists. Those paragraphs were very brief, each amounting to no more than a dozen words. It made no mention of important fungus areas, threats to fungi, or deficiencies in legal protection for fungi. Finland's report contained separate short paragraphs exclusively devoted to fungal conservation, and specifically to important fungus areas. It also briefly mentioned fungal checklists and red lists, threats to fungi, *in situ* conservation of fungi, and education, but all in paragraphs shared with plant conservation. In Lithuania's report there was one short paragraph considering conservation threats which was devoted exclusively to fungi. Fungal checklists, red lists, *in situ* conservation and legal protection were also considered, but all jointly with animals and plants. Serbia's report briefly mentioned the need for inventories and red lists for fungi, but again jointly with animals and plants. Sweden's report, in a paragraph devoted to mycorrhizal fungi, commented very briefly on the specific threat of nitrogen deposition and, in a paragraph devoted to reindeer herding, commented on the threat to grazed lichens from climate change. It also acknowledged that some groups of fungi (along with other groups of animals and plants) contain particularly large numbers of threatened species, and mentioned fungal inventories and red lists (also along with animals and plants). The report from the UK had two entire sections devoted to fungi and lichens. It considered the few existing fungal checklists and red lists, and identified the need to complete inventories and conservation assessments for fungi. There was no mention of important fungal areas and very little discussion of legal protection for fungi.

CONSIDERATION OF FUNGAL HABITATS AND ROLES OF FUNGI

Of the example fungal habitats and roles listed in question 4 above, the following were considered in none of the documents: dung fungi, endobionts, freshwater fungi, fungi on man-made products, fungi on naturally occurring inanimate substrata or marine fungi.

None of the documents expressed any concern that so little is known about the likely impact of climate change, pollution and habitat loss on fungi in these habitats, or of the knock-on effects of those impacts. Very few reports considered fungi as decomposers, as mycorrhizal symbionts, or as animal or plant parasites. The importance of fungi as decomposers was briefly noted in the report from Bosnia and Herzegovina, while the report of Sweden referred briefly to the importance of decomposing wood as a fungal habitat. The report from Lithuania very briefly noted that habitat fragmentation, intense economic activities and pollution constitute a threat to parasitic fungi. The report from Sweden mentioned mycorrhizal fungi once, very briefly. The UK report, and some others, considered parasitic fungi, but only as something undesirable or potentially undesirable, not as organisms worthy of conservation.

No report expressed concern that so little is known about threats to such habitats or about the knock-on effect of such threats. No other fungal habitats or roles were discussed.

RECOGNITION OF THE KNOWLEDGE GAP, AND PLANS TO DEAL WITH THAT PROBLEM

Of the 44 countries reviewed, 33 showed no awareness at all in their reports that there was a problem with lack of knowledge about fungi. Not surprisingly therefore they also contained no discussion at all about resources needed for fungal conservation. Of the remaining 11, Belgium, Bulgaria, Cyprus, Denmark, Germany, Serbia and Slovakia all made some mention of the problem, usually very brief, but gave no consideration to how the knowledge gap might be resolved. They also contained no discussion about resources needed for fungal conservation.

Austria's report acknowledged the existence of a knowledge gap, and stated (without providing any further evidence) that substantial progress had been made in bridging it through a start in research and documentation. Elsewhere, however, the same report admitted that "assessment of fungi is feasible only for a selected group of rather well-known species", a statement which made the earlier claim of progress look rather unconvincing. Resources for fungal conservation were not discussed. The Polish report acknowledged the knowledge gap, and stated very briefly (without any discussion of resources) that some effort is being made to include identification of fungi (along with animals and plants) in the country's national curriculum. Finland's report recognized the knowledge gap, and the challenge of a lack of fungal specialists and of other personnel responsible for consultation and negotiation with landowners.

Its report stated that environmental literacy (in particular identification of local animals, fungi and plants) constitutes a core content of the national school curriculum (with resources from the country's education budget implied but not explicitly mentioned).

It also reported that the country's Natural History Museum has included information about fungi in its new exhibitions. No other resources were discussed. The UK's report acknowledged the knowledge gap and the need for more mycological (and botanical and zoological) expertise, and on page 111 of the document described its plan to address the problem.

Unfortunately, on closer reading, the plan is largely a description of high minded but unfunded strategies and initiatives by public institutions like museums and botanic gardens, or learned societies and charitable organizations, and most of those activities had already come to an end by the time the report was being written. There was no mention at all of any real funding.

BALANCED TREATMENT?

In every document there was detailed and separate discussion of the conservation status and strategies for protection of amphibians, birds, fish, mammals and reptiles. These are all rather small vertebrate groups at or near the taxonomic rank of class.

In comparison, very few of the documents contained any structured consideration of fungi at taxonomic ranks below kingdom level (for example the *Ascomycota*, *Basidiomycota* and other fungal *phyla*, some very substantial in size).

Attention to such groups, even though at a significantly higher taxonomic level than class (as used for vertebrates), was almost entirely lacking.

The report from Belgium mentioned that checklists of ascomycetes and basidiomycetes were in preparation in Flanders.

The reports for Bosnia and Herzegovina, Lithuania and Montenegro gave statistics for species numbers in various fungal groups at phylum or class level, none of them comprehensive. The UK's report mentioned *Ascomycota*, *Basidiomycota*, *Chytridiomycota*, *Glomeromycota* and *Zygomycota*, but only in the context of checklists, most of which did not exist.

No country considered the *Basidiomycota*, for example, as important mycorrhizal fungi (*e.g. Agaricales*), or as wood decay fungi (*e.g. Polyporales*), or as leaf parasites (the rusts), or as flower parasites (the smuts), and no country considered the *Ascomycota* in any of the huge range of habitats they occupy.

DISCUSSION

The 44 action plans and reports examined in this chapter represent the most recent statements to the CBD by European governments of what they are doing about biodiversity conservation. Collectively, these documents contain very little which is meaningful about fungal conservation. The findings reported in the previous paragraphs will certainly not surprise mycologists. The Rio Convention has not served fungi well, and fungi are already frequently referred to as the "Orphans of Rio" [www.fungal-conservation.org/blogs/orphans-of-rio.pdf]. The findings are also unlikely to surprise any biologist or conservation expert. They are, however, truly shocking. From the point of view of fungal conservation, only Finland's report was anywhere near adequate, and even that provided consideration and coverage of the fungi which left a lot to be desired.

All of the other documents examined were worse, usually far worse. Almost all of them were dreadful. On the basis of their own reports, it is clear that most European governments have only the most sketchy idea of what fungi are, and no idea of why it might be important to conserve them. Where fungi were mentioned, the over-riding impression of this reviewer was that, in most cases, at a late stage in the drafting, someone had gone through the text searching for "plants and animals", and had added "and fungi", in the belief that this somehow resulted in a document with satisfactory treatment of fungal conservation. In most cases, the idea that fungi might have their own special needs seems never to have occurred.

Reviewing these texts was a depressing experience. It might have been hoped that Europe, with all its wealth and resources, and its long history of nature conservation movements, would have been a continent where CBD reports really did address biodiversity conservation properly. But time and time again, shocking surprises were encountered. How can countries so dependent on forestry as Finland, Norway and Sweden produce reports which show so little awareness of the importance of mycorrhizal fungi, tree endobionts or litter decomposing species? How did France, Italy and Spain all manage to produce biodiversity conservation plans which do not consider protection of truffles? How can countries faced with climate change ignore the fungal component of soils? How can countries which have seen a catastrophic decline in the expertise and resources necessary for biodiversity conservation be so indifferent to the problems which that decline has caused? How can it be sensible to pretend that the issue of proper funding for necessary work does not exist? How can any of the countries imagine that plans to conserve the producers (plants) and consumers (animals) will work correctly without any thought for the recyclers? How can they assume so complacently that fungi will somehow magically be protected against climate change, habitat loss and pollution?

It was evident throughout that fungi are not alone in getting such inadequate treatment. Similar searches could be made for arthropods, echinoderms, molluscs, nematodes and a whole army of other invertebrate animals, or for algae, bacteria or protozoans, or for many plant groups, and similar distressing results would have been obtained. These CBD strategies and reports are not fit for purpose for fungi, or for most of the animal kingdom, or for most of the other species on this planet. That raises the interesting question: how are these documents audited? The CBD is, after all, just like a gigantic international project, and in any project, it is normal for plans and reports to be checked, subjected to critical comment and otherwise audited by independent external reviewers. Is there some system by which plans and reports coming in to the CBD receive such critical scrutiny? If there is, the system doesn't seem to be working, and needs to be changed. If there isn't, then it's about time a system was put in place.

The appalling neglect of the fungi by the CBD is something which should worry not just mycologists, but also botanists, zoologists, everyone involved in the conservation movement in general, the educated public at large, and certainly the governments who were responsible for producing such unsatisfactory work. If this vastly important kingdom of organisms continues to be overlooked, then conservation of animals and plants will also certainly be impaired. The key question is, "what can be done to change this unsatisfactory situation?" The answer lies in science, infrastructure, education and politics, and the only way to deliver that answer is through a consistent message that fungi are not animals or plants but are something different which is very special and very precious.

Science

The fungus which forms the theme of the present ebook is a good example of how science works to promote fungal conservation, and the review of CBD documents in this chapter only makes more clear how great an achievement it was to get this species red listed. A lot of the science which has stimulated the fungal conservation movement has come from Europe, from the early work with air pollution and lichens, through the pioneering studies of the 1980s showing major declines in larger fungi populations [1], to more recent work demonstrating responses by fungi to climate change [2].

The landmark book *Fungal Conservation. Issues and Solutions* [3] was produced in Europe and strongly influenced by European thought and research, and Europe continues to produce an impressive array of scientific evidence underpinning fungal conservation.

European mycology has contributed to the *IUCN'S Sampled Red List Index* project [www.zsl.org/science/research-projects/ia-iucn-redlist,1161,AR.html] with 1500 baseline evaluations of ascomycetes [www.cybertruffle.org.uk/redlidat], and an excellent very recent example of science supporting fungal conservation is the evaluation of British lichens and lichenicolous fungi by Woods & Coppins (2012) [4].

The role of digitized information, particularly where it is available on-line, has also been important. In addition to general resources for biodiversity, such as the *Global Biodiversity Information Facility* [www.gbif.org], the *Biodiversity Heritage Library* [www.biodiversitylibrary.org] and the *Encyclopedia of Life* [http://eol.org], there are many specialist fungal websites based in Europe. *IndexFungorum*, the fungal nomenclator [www.indexfungorum.org], and *Mycobank* [www.mycobank.org], which is similar but also provides descriptions of species and other information, are key resources.

Cyberliber [www.cybertruffle.org.uk/cyberliber] provides free and open access to nearly 400,000 pages of mycological literature, and there are many enormous databases accessible on the internet which provide information about when, where and on what fungi occur including, for example, *Cybertruffle's Robigalia* [www.cybertruffle.org.uk/robigalia/eng] and the *Fungus Records Database of Britain and Ireland* [www.fieldmycology.net/FRDBI/FRDBI.asp]. These (all based in Europe) are making a huge contribution to fungal conservation simply by providing the basic information necessary for the job: their databases have been the foundation of much of the recent work producing printed checklists, for example, Minter and Dudka (1996) [5] and Legon and Henrici (2005) [6], and similar on-line digital products [for example www.cybertruffle.org.uk/ukrafung/eng]. Tellingly, many of them receive no government funding.

Infrastructure

More than 50 years ago, there were already publications showing that some lichen-forming fungi were threatened, and societies devoted to lichens started to consider action to deal with the problem. For non-lichen-forming fungi, the *IX Congress of European Mycologists in Oslo* in 1985 saw establishment of the *European Committee for Fungal Conservation*, which was perhaps the first organized body in the world to address fungal conservation. In 2003, that became the conservation arm of the *European Mycological Association*. By the end of 2008, similar conservation groups had been formed by the continental level learned societies for mycology in Africa, Asia, Australasia, North America and South America. Various national mycological societies also by this point had conservation groups. To date, all of these have restricted their activities to education and promoting the science underlying fungal conservation.

Up to 2008, the only bodies dealing with the political side of fungal conservation were two specialist groups for fungi in the *Species Survival Commission* of the IUCN (one for lichen-forming species, the other for all other fungi).

In that year, the number of IUCN fungal specialist groups was increased to five (three of them based in Europe), and for the first time they were clearly distinguished from the plant specialist groups.

There was still, however, no society anywhere in the world exclusively dedicated to fungal conservation. After preparatory meetings in Córdoba (Spain, 2007) and

Whitby (England, 2009), that gap was filled at a meeting in Edinburgh (Scotland, 2010) with the foundation of the *International Society for Fungal Conservation* [www.fungal-conservation.org], a society which explicitly recognizes the political element of conservation.

All of that was progress. But compared with animal and plant conservation, the infrastructure for fungal conservation remains very under-developed, and the resources available to fungal conservationists are vanishingly small.

There are no national fungal conservation societies anywhere. There are no organizations like *Botanic Gardens Conservation International* or the *Royal Society for Protection of Birds*, and most of the big international conservation NGOs, and most major biodiversity institutions such as natural history museums are as little aware of the importance of fungi as the governments which wrote those CBD reports, with the result that they rarely have staff who can deal with fungi in general and their conservation in particular.

To bring about change, mycologists need to support the fungal conservation infrastructure which already exists, making sure that it functions.

They also need to develop that infrastructure, so that there are continental, national and local level societies for fungal conservation, and further societies specializing in conservation of particular groups, particularly iconic organisms like truffles or wax caps.

From that platform, mycologists then need to ensure that fungi have a voice in conservation at the same level as the voices for animals and plants.

Wherever there is a committee or group populated by botanists and zoologists (for example an advisory committee for allocating research funds for conservation), there ought also to be mycologists present.

Education

At all levels of society, awareness of the importance of fungi remains very low. In the UK, for example, fungi do not feature in the national curriculum for schools, with the result that children leave school having finished secondary education with no understanding of what fungi are. For those going on to university education, there are currently no courses exclusively dedicated to mycology. All of these children become voters. Some become politicians, or civil servants, or directors of museums or other institutions. Some even become rich. It is a disaster for fungal conservation if they do not understand that fungi are important.

As a result, mycologists need to promote education about fungi at all levels. That means getting mycology into schools, colleges and universities.

It also means challenging very common assumptions which have resulted from the poor education in biodiversity prevalent at the moment. The almost universal practice of using "fauna and flora" or "animals and plants" as a lazy shorthand for all biodiversity is one of the most damaging. The CBD documents evaluated above were full of these phrases, but the practice extends far more widely and has the effect of an intellectual straitjacket restricting the users' perception. A classic example is the logo of the United Nations 2010-2020 decade on biodiversity, which shows only animals and plants (Fig. **1**).

United Nations Decade on Biodiversity

Figure 1: The official logo of the United Nations Decade on Biodiversity.

Wherever these misleading assumptions are seen, mycologists need to draw fungi to the attention of those responsible.

The *International Society for Fungal Conservation* has identified the pages of Wikipedia as one key area for action. Everybody likes to criticise Wikipedia, but everybody uses it.

Society members are encouraged to edit pages which deal with biodiversity to make sure fungi are properly represented. Weaknesses can also be turned to strengths.

The 2010-2020 decade on biodiversity logo mentioned above is available for use by conservation organizations and, when positioned adjacent to suitable additional text, is very appropriate for delivering the message of fungal conservation (Fig. **2**).

Figure 2: An example of how to use the UN Decade on Biodiversity logo in a way which promotes awareness of fungal diversity.

Another key area of education lies in making our closest friends, the botanists and zoologists, aware of the important role they can play in supporting fungal conservation [www.fungal-conservation.org/blogs/message-to-botanists-and-zoologists.pdf].

In the UK, most recent governmental review of taxonomy interviewed various scientists, but not a single mycologist.

A senior botanist was asked for an opinion about the state of mycology. The reply was that the state of mycology was catastrophic.

This was accurate and truthful, but an even better reply would have been, "I am a botanist; to find out about fungi, you need to ask a mycologist".

That would have given mycology a voice.

Mycologists need to remind botanists and zoologists that they do not represent mycology, and to persuade them that, whenever asked about fungi, the correct response is to redirect the questioner to someone who does represent the subject.

We also need to raise awareness of fungi in all areas of civil society where there is an interest in conservation. That means engaging local, provincial, national and international societies, government agencies and environmental ministries. Where fungi are overlooked, the deficiency needs to be pointed out. Where fungi are recognized, praise is needed, and encouragement to allocate resources to fungal conservation.

Politics

For all their ideals, fungal conservationists have to live in the real world, where politics cannot be ignored. Getting the CBD to deliver genuine biodiversity conservation - that means protection of fungi, invertebrate animals and all the other neglected groups - is clearly not going to happen overnight. But it will never happen at all as long as the pretence is maintained that the present system is satisfactory. Keeping silent about the deficiencies of the CBD is therefore simply not an option. The problem mycologists face is that of getting heard: there is a glass ceiling between us and those who can make these necessary changes real.

Politicians do not write documents for the CBD, nor do their minions. Most never see these reports, and it's unlikely that many read them. The documents are compiled in the ministries, and those responsible for compiling them consult safe and established sources: senior staff in governmental environment agencies and research councils, directors of natural history museums, keepers of national botanic gardens and so on. Those sources then delegate, and by the time the job has filtered down to the level of someone who actually does the writing, the remit has become ossified and there is no opportunity for thinking outside the box. The quality of the resulting reports should not come as a surprise.

Fungal conservationists need to find some way past all of this. The challenges in the UK are as good an example as any. The country's oldest and most prestigious biological society, the Linnean, until very recently had an internal structure which was still based on the dichotomy of fauna and flora. As a result of pressure from mycologists, that structure is now being modified. The national Natural History Museum still tends to present biodiversity in terms of animals and plants. The Royal Botanic Gardens, the country's premier location for plant taxonomy, also houses the country's equally important national fungus collection but this is never publicized and gets only a tiny portion of the institution's resources. These taxonomic institutions are conservative and slow to change, and their failure to move with the times makes them appear shockingly out-of-date when it comes to recognizing fungi. Unfortunately, they are also the key organizations which need to change if fungal conservation is to make any progress at all.

Fortunately, there are other equally prestigious bodies which are more nimble. The IUCN, the world's leading conservation NGO, has distinguished itself by its willingness to embrace change, and that is now reflected throughout the organization. In the IUCN, policy is determined by a special congress held every four years, the most recent being in South Korea in September 2012. At that congress, for the first time ever, delegates considered a motion relating to fungi. In brief, that motion proposed that the IUCN should formally recognize that fungi are just as important as animals and plants, and should in future prioritize their conservation activities accordingly. The motion was passed with an overwhelming majority. If the IUCN can do it, why not the Linnean Society, or the Natural History Museum, or the Royal Botanic Gardens?

To conserve fungi, mycologists need to learn how to work in a political environment. Campaigns need to be attractive and effective. Politicians will only change policies if there are votes in it. Their advisors must also be targeted. The national focus points for the CBD in each country are listed on the CBD website, with their postal addresses and, very often, their e-mail addresses [www.cbd.int/information/nfp.shtml].

Writing to them pointing out the importance of fungal conservation might be a good place to start.

CONCLUSIONS

European governments urgently need to change their attitude towards biodiversity conservation. Their coverage of fungi in the CBD documents reviewed in this chapter is almost universally inadequate. The inadequacy is inexcusable. Fungi have been recognized as a totally separate biological kingdom for well over 40 years, and for at least 20 years there has been widespread agreement that this kingdom is likely to include more species than there are in the plant kingdom [7]. The inadequacy is also stupid. The fungi are not a small and insignificant group of organisms, they are a huge assembly of recyclers without which life on this planet, as we know it, would be unsustainable. Ignoring them, or pretending they do not exist is not an intelligent option. Lastly, the inadequacy is deeply worrying. The main threats to biodiversity are climate change, habitat loss and pollution. Fungi have no special protection against these drivers of biodiversity loss. They are just as endangered as animals and plants. Neglecting them simply moves our own species further up the endangered list. Mycologists face a big challenge in educating governments to value fungi properly.

ACKNOWLEDGEMENT

Declared none.

CONFLICT OF INTEREST

The author(s) confirm that this chapter content has no conflict of interest.

REFERENCES

[1]   Arnolds E. The changing macromycete flora in The Netherlands. T Brit Mycol Soc 1988; 90(3): 391-406.

[2]   Kauserud H, Heegaard E, Semenov MA, *et al.* Climate change and spring-fruiting fungi. Proc R Soc B 2010; 277(1685): 1169-77.

[3]   Moore D, Nauta MM, Evans SE, Rotheroe M, Eds. Fungal Conservation. Issues and Solutions. Cambridge University Press: Cambridge 2001; i-x, pp. 1-262.

[4]   Woods RG, Coppins BJ. A Conservation Evaluation of British Lichens and Lichenicolous Fungi. Species Status 13: Joint Nature Conservation Committee, Peterborough: UK 2012; i-v, pp. 1-155.

[5]   Minter DW, Dudka IO. Fungi of Ukraine. A Preliminary Checklist. International Mycological Institute, Egham, Surrey, UK & M.G. Kholodny Institute of Botany, Kiev: Ukraine 1996; pp. 361.

[6]   Legon NW, Henrici A. Checklist of the British and Irish Basidiomycota. Royal Botanic Gardens, Kew: UK 2005; xvii, pp. 517.

[7]   Hawksworth DL. The fungal dimension of biodiversity: magnitude, significance, and conservation. Mycol Res 1991; 95(6): 641-55.

APPENDIX 1

Comparative performance of different European countries in fungal conservation on the basis of their most recent CBD reports:

Iceland   No mention of fungi.

Luxembourg   No mention of fungi.

Monaco   No mention of fungi.

Spain   No mention of fungi.

"Deficient in Four Aspects"

Albania   Fungi mentioned, but not clearly distinguished from plants; lichens treated as separate from fungi.

Belarus   Fungi mentioned (as "mushrooms"); not clearly distinguished from animals, micro-organisms or plants; lichens treated as "lower plants".

Belgium   Fungi mentioned; lichens mentioned; treated as plants; some recognition of knowledge gap, but no plans to address the problem.

Cyprus   Fungi mentioned (as "mushrooms"); lichens listed under flora; some recognition of knowledge gap.

Czech Republic Fungi mentioned; mostly treated as plants; some minimal recognition of threats and need for protection, but no plans to deal with them.

European Union Fungi mentioned; no evidence of awareness that they might be different from animals or plants; no consideration of red lists, threats or fungal habitats; no acknowledgement of knowledge gap.

France   Fungi mentioned, but separate status not explicit.

FYROM   Fungi mentioned; lichens listed separately from fungi; both listed separately from animals, micro-organisms and plants.

Greece   Lichens mentioned once; no indication of any understanding of their status.

Hungary   Fungi and lichens mentioned and clearly and consistently distinguished from animals and plants, but also treated as distinct from each other.

Italy   Fungi mentioned; lichens listed separately from fungi, and both included in "flora".

Latvia   Fungi mentioned; clearly and consistently distinguished from animals, micro-organisms and plants, but lichens confused with plants.

Liechtenstein   Fungi mentioned (as "mushrooms"); some evidence that they are considered separate from plants; some evidence that there is interest in their protection.

Malta   Fungi mentioned; confused with micro-organisms and plants; lichens mentioned but listed as "lower plants".

Moldova   Fungi mentioned (as "mushrooms"), largely in the context of harvesting, and treated as plants; lichens mentioned once, and treated as plants.

The Netherlands   Fungi mentioned (as "macrofungi"), and listed as plants; lichens mentioned separately from fungi, and also listed as plants.

Norway   Fungi mentioned, but only in the context of forest tree disease or (as "mushrooms") in the context of sustainable harvesting; lichens mentioned in the context of pollution monitoring.

Portugal	Fungi mentioned; lichens mentioned; described as plants.
Romania	Fungi and lichens mentioned, but treated as plants.
Slovakia	Fungi and lichens mentioned, but treated as plants; some acknowledgement of knowledge gap, but not plans to address it.
Slovenia	Fungi mentioned, with some indication of being taken into account in plans, but status not clear (mentioned more than once as part of plant kingdom, and once as microbial life).
Turkey	Fungi mentioned; lichens mentioned; both described as plants.
Ukraine	Fungi mentioned; confused with micro-organisms and plants; lichens mentioned; listed separately from fungi, confused with plants.

"Deficient in Three Aspects"

Austria	Fungi mentioned; generally but not consistently distinguished from plants; lichens listed as plants; knowledge gap for fungi acknowledged and some plans in place to address that gap.
Bulgaria	Fungi mentioned in a conservation context; not consistently distinguished from plants; lichens listed as plants; some strategic consideration of fungal conservation (fungi explicitly included in national red listing); need to improve informational resources about fungi acknowledged.
Denmark	Fungi mentioned; lichens mentioned; generally recognized as distinct, but sometimes listed as plants; some strategic consideration of fungal conservation (some recognition of threats); some fungal habitats recognized; some recognition of the knowledge gap.
Estonia	Fungi and lichens mentioned and recognized as distinct from animals, micro-organisms, plants and from each other; some

strategic consideration of fungal conservation (fungal and lichen red lists mentioned).

Germany   Fungi mentioned; lichens not mentioned; fungi distinguished from animals, micro-organisms and plants, but not consistently (also listed in the context of "flora"); knowledge gap acknowledged for fungi.

Ireland   Fungi mentioned; fungi and lichens distinguished from animals, micro-organisms and plants, but listed separately from each other; a lichen red list in preparation for Ireland mentioned.

Montenegro Fungi mentioned; generally clearly distinguished from animals, micro-organisms and plants, but lichens sometimes confused with plants (*e.g.* listed with bryophytes); important fungal areas mentioned; legislation relating to fungal conservation mentioned.

Poland   Fungi and lichens mentioned; usually distinguished as separate, but sometimes treated as plants; some minimal consideration of threats; some awareness of the knowledge gap, and some minimal evidence of plans to address the problem.

Russia   Fungi and lichens mentioned; listed separately from animals, plants and from each other.

Switzerland Fungi mentioned; treated as distinct from animals, micro-organisms and plants; lichens treated as distinct from fungi, micro-organisms and plants; existence of red lists for some fungi and for lichen-forming fungi provides some evidence of strategic consideration.

"Deficient in Two Aspects"

Bosnia and   Fungi mentioned; recognized as a separate biological kingdom;
Herzegovina lichens recognized as fungi; fungi clearly taken into account in

conservation planning; some recognition of different fungal habitats, but incomplete.

Croatia   Fungi mentioned and clearly distinguished from animals, micro-organisms and plants; some strategic consideration of fungal conservation (reference to red lists and legal protection; some planning for sustainable use of fungi).

Lithuania   Fungi mentioned and consistently treated as distinct from animals and plants (although included in the country's contribution to the Global Strategy for Plant Conservation); lichens mentioned and consistently treated as fungi; some strategic consideration of fungal conservation (red listing for fungi mentioned, and some awareness of threats).

Sweden   Fungi mentioned and consistently treated as distinct from animals and plants (although included in the country's contribution to the Global Strategy for Plant Conservation); lichens mentioned but treated separately from fungi; some strategic consideration of fungal conservation (red lists mentioned, with some plans to address threats); some consideration of fungal habitats.

"Deficient in One Aspect"

Finland   Fungi mentioned; consistently distinguished from plants (although included in the country's contribution to the Global Strategy for Plant Conservation); lichens consistently recognized as fungi; some strategic consideration of fungal conservation (frequent mention of red lists and some other tools); some fungal habitats recognized; knowledge gap acknowledged and some plans in place to deal with the problem.

Serbia   Fungi mentioned and present on logo; clearly recognized as a separate biological kingdom; lichens recognized as fungi; some strategic consideration of fungal conservation (mushrooms considered as part of policy on foraging; truffles considered as part of forest resources); some fungal habitats recognized; some attention given to dealing with knowledge gaps in fungi.

UK   Fungi and lichens mentioned, and generally recognized as distinct from animals and plants (although included in the country's contribution to the Global Strategy for Plant Conservation); some strategic consideration of fungal conservation (red listing and threats mentioned); some fungal habitats mentioned; knowledge gap acknowledged, and some discussion of resources for dealing with the problem.

CHAPTER 2

The Genus *Pleurotus* (Fr.) P. Kumm. (Pleurotaceae) in Europe

Georgios I. Zervakis[*] and Elias Polemis

Agricultural University of Athens, Laboratory of General and Agricultural Microbiology, Iera Odos 75, 11855 Athens, Greece

Abstract: The genus *Pleurotus* comprises ca. 30 species and subspecific taxa of edible mushrooms with a world-wide distribution. Most of them are cultivated on a large range of agricultural and forestry residues and by-products providing a relatively cheap food of high dietetic value through rather simple solid-state fermentation processes. In addition, *Pleurotus* biomass demonstrates significant medicinal effects and its bioactive compounds (mainly polysaccharides) possess antibiotic, antitumor, hypocholesterolemic and immunomodulation properties. One of the most important aspects related with the exploitation of *Pleurotus* fungi is that they are powerful lignin decomposers and hence they are used as potent biodegraders of numerous organic pollutants, xenobiotics and industrial wastes. However, all such biotechnological applications are tightly linked with the isolation, identification, evaluation and improvement of the respective genetic resources. Assessments of *Pleurotus* diversity in Europe in conjunction with biochemical, molecular and compatibility studies revealed the existence of eight species, *i.e. P. calyptratus*, *P. cornucopiae*, *P. dryinus*, *P. eryngii*, *P. fuscosquamulosus*, *P. nebrodensis*, *P. ostreatus* and *P. pulmonarius*, which are described in detail (anatomy, ecology and distribution). Furthermore, *P. abieticola* and *P. opuntiae* are two additional species reported to occur in Europe, albeit infrequent to very rare; the former was isolated from east Russia whereas the latter in Mediterranean Europe. A synthesis of available information on *Pleurotus* systematics is also presented and discussed.

Keywords: Agaricales, Apiaceae, Basidiomycota, Biodegradation, Cultivation, Decomposers, Distribution, Europe, Evolution, Fungal biotechnology, Genetic resources, Habitat, Medicinal mushrooms, Molecular analysis, Organoleptic properties, Phylogeny, Pleurotaceae, Pleurotus, Systematics, White-rot fungi.

INTRODUCTION

The genus *Pleurotus* (Fr.) P. Kumm. (Pleurotaceae, Agaricales) constitutes an

*Address correspondence to Georgios I. Zervakis: Agricultural University of Athens, Laboratory of General and Agricultural Microbiology, Iera Odos 75, 11855 Athens, Greece; Tel: +302105294341; E-mail: zervakis@aua.gr

Maria Letizia Gargano, Georgios I. Zervakis and Giuseppe Venturella (Eds)

important group of edible and medicinal mushroom species, which have a cosmopolitan distribution, and usually grow as saprotrophs on dead wood and plant residues. *Pleurotus* spp. are classified within the physiological grouping of white-rot fungi, *i.e.* fungi that are capable of extensively degrading lignocellulosic substrates in order to access polysaccharides locked in lignin-carbohydrate complexes in order to provide an energy source which other organisms cannot reach [1]. Such fungi possess the additional and particular ability for delignification by being able to synthesize the relevant hydrolytic (cellulases and hemicellulases) and oxidative (ligninolytic) extracellular enzymes required to degrade the major components of the substrate into low-molecular-weight compounds that can be assimilated for their nutrition [2,3]. The cultivation of *Pleurotus* spp. (notably *P. ostreatus*, *P. eryngii*, and *P. pulmonarius* in temperate regions; *P. djamor* and *P. cystidiosus*/*P. abalonus* in tropical and subtropical areas) is an economically important agro-industrial activity practiced on a wide variety of lignocellulosic wastes. *Pleurotus* (or oyster) mushrooms production accounts for ca. 15% of the world total, *i.e.* about one million tons per year [4]. Their cultivation is a relatively simple process, which demands heat pasteurization of the substrate, but without prior composting or pretreatment of the initial materials. Furthermore, incubation and fruiting does not necessarily require expensive infrastructure, it presents few obstacles as far as pests and diseases are concerned, and in most cases it does not need the amendment of a casing layer. The principal component used for the production of the *Pleurotus* substrate is usually cereal straw, which is shredded, wetted with water and often supplemented with nitrogen-rich media (*e.g.* meals and flours deriving from leguminous crops). However, there is an ever-growing and world-wide tendency to exploit the *Pleurotus* lignocellulosic enzymes for valorizing a large array of agricultural and forestry wastes (*e.g.* wood chips, sawdust, coffee pulp, corn cobs, coconut palm wastes, sugarcane bagasse, cotton waste, olive press cake, orange peels, grape stalks, *etc.*) as alternative substrates for oyster-mushroom cultivation [5-16]. *Pleurotus* mushrooms possess fine organoleptic properties (unique flavor and aroma), and they are rich in carbohydrates (3-28% f.w.), dietary fiber (3-32% d.w.), proteins (10-30% d.w.), essential amino acids (especially alanine, glutamic acid, and glutamine: 25-35% d.w.), vitamins (C, A, B2, B1, D and niacin), minerals (K, P, Mg, Ca, Na, Fe, Se, Zn, Cu, Mn), and lipids (3-5% d.w.) [17]. Furthermore, *Pleurotus* spp. are promising as medicinal mushrooms exhibiting activity against various diseases [18,19]. Several bioactive molecules have been identified in these mushrooms, such as antioxidants, dietary fibres, polysaccharides (*e.g.* β-D-glucans of heterogeneous molecular weights), lectins and glycopeptides, which were shown to modulate the immune system, inhibit tumour growth and inflammation, have hypoglycaemic and antithrombotic activities, lower blood lipid concentrations, prevent high blood pressure and atherosclerosis, and present antimicrobial and other health-promoting activities [17,19-26]. Except of food, *Pleurotus* spp. have been used for the production of good quality fodder by enriching the initial wastes with protein through mycelium growth (biomass supplementation) and by improving at the same time, digestibility by preferentially consuming lignin and hemicellulose and leaving cellulose fairly intact as an energy source for ruminants [3,27]. In this way, different types of wastes (*e.g.* wheat straw, cotton stalks, *etc.*) were upgraded and rendered suitable for animal feed [28-32]. However, one of the most interesting applications of *Pleurotus* spp. is related to the exploitation of their unique ligninolytic system as a tool for the biotransformation/biodegradation of industrial effluents, and of wastewaters/residues deriving from agricultural, forestry and agro- industrial activities. Three ligninolytic enzyme families have been characterized: manganese-dependent peroxidase, versatile peroxidase and laccase [33,34]. Hence, *Pleurotus* fungi have been successfully used for the biodegradation of polycyclic aromatic hydrocarbons (PAHs) [35,36], bioremediation of soils polluted with PAHs [37,38], mineralization of DDT [39,40], decolorization of dyes from textile and other industries [41,42], detoxification of agro-industrial effluents [43,44], biodegradation of organophosphorus insecticides [45,46], *etc.* The association of *Pleurotus* fungi with such a large array of important commercial and industrial applications demands a clear image of the relationships among genetically-related populations, which in turn will provide the basis for controlled breeding and successful exploitation of such resources so that suitable and correctly identified strains in addition to well-adapted biotechnological methodologies could be rendered available.

PLEUROTUS SYSTEMATICS AND PHYLOGENY

The genus *Pleurotus* (Fr.) Kummer comprises gilled mushrooms with pleurotoid (*i.e.* with eccentric or lateral stem or laterally attached and sessile) basidiomata. Together with *Hohenbuehelia* S. Schulz., they are the only genera of the Agaricales known to attack and consume living nematodes [47-49]. Thorn and

Barron [47] argued that this shared biological habit merited their inclusion in one family, the Pleurotaceae; the lack of this habit was later used as support for the exclusion of species previously included in the genus *Pleurotus* [50]. Circumscription of the genus *Pleurotus* has been difficult and controversial. Each of the taxonomic systems of several authors [51-54] presented a different delimitation (and, in some cases, type concept) of these genera and their putative allies [50,55]. The results of several phylogenetic studies [56-58] clearly indicated that the monophyletic genera *Pleurotus* and *Hohenbuehelia* are included within the agaricalean pleurotoid clades and together form a monophyletic Pleurotaceae. Therefore, the biological character of nematophagy among pleurotoid fungi is consistent with phylogenies based on rDNA sequences [55,58]. The family can thus be defined in part on the basis of biotrophy, *i.e.* capture and consumption of living nematodes [58]. In addition, species of *Pleurotus* can be defined (and hence differentiated from *Hohenbuehelia*) by the absence of a gelatinous zone (although some species have a gelatinized pileipellis), the absence of metuloid cystidia, the absence of a *Nematoctonus* anamorph, and the production of liquid microdroplets containing the nematotoxin *trans*-decenedioic acid [59] on secretory appendages, which are located on fine, tapering pegs scattered on aerial somatic hyphae. There has been significant controversy in the past, as concerns the systematics of *Pleurotus* spp., especially regarding relationships among several "true" taxa or wrongly applied taxonomic names [60-63]. These taxonomic ambiguities arose because of initial misidentifications, the effect of environmental conditions on morphological characters, and the limited number of suitable mating studies which could be combined with robust ecophysiological features.

For resolving systematics, phylogeny and evolution of *Pleurotus* spp., the implementation of integrated approaches were necessary, and this was achieved through the combined use of various methodologies. The interpretation of morphological, physiological, and ecological data in conjunction with the outcome of compatibility, isozyme and molecular studies provided the much needed framework for drawing accurate conclusions for a several *Pleurotus* taxa. Recent investigations using molecular criteria to infer *Pleurotus* systematics, phylogeny and evolution also shed light on the perplexed mating behavior of several strains within the genus [64,65]. In fact, instead of negating the utility of mating studies to understand species concepts, it is important to remember that partial compatibility provide valuable data on the evolutionary dynamics of reproductive isolation and speciation. It is therefore essential to combine such approaches with morphological/physiological data and DNA analyses to come up with "natural" classification systems in *Pleurotus*.

PLEUROTUS TAXA OCCURRING IN EUROPE - DESCRIPTIONS AND COMMENTS

All morphological descriptions that follow are mainly based on the outcome of a pertinent study by Zervakis and Balis [62].

Pleurotus calyptratus (Lindblad) Sacc

Description: Pileus (2.5–) 4.0–11 cm wide, initially convex, smooth, semicircular to kidney-shaped, at maturity circular to cell-like, mostly flat, smooth, silky-fibrillose, slightly viscid, with broad, initially inrolled then sinuate margin, often with remains of veil; hygrophanous grayish blue, grayish brown to warm brown, later beige-buff to light beige, cream to almost white. Lamellae entire, thin, crowded, decurrent to the stipe top, without evident anastomoses, with several lamellulae often forked, at first white then cream to light beige. Stipe absent, rudimentary or lateral. Existence of veil which extends from the margin to the base of pileus, white to cream. Odour pleasant, fruity, farinaceous or indistict. Taste mild, similar. Flesh white to cream, solid and thick. Spore print (dry) white to cream.

Basidiospores (8.5–)10.5 – 15.5 (–17.0) x (3.0–) 4.0 – 5 (–6.5) µm, thin-walled, cylindric to elliptic with a small broad apiculus. Basidia 35 – 45 x 8.0 – 11.0 µm, 4-spored, cylindric, thin-walled, with sterigmata up to 5 µm long. Cystidia lacking. Pileus trama dimitic with tightly interwoven, thin-walled and sclerified generative hyphae 4.0 – 7.0 µm wide, and skeletal hyphae 4.0 – 6.5 µm wide. Pileipellis 120 – 140 µm thick, with irregularly interwoven, generative hyphae. Lamellar trama dimitic with generative hyphae 5.0 – 6.0 µm wide, sclerified thick-walled hyphae, 6.5 – 10.0 µm wide, and skeletal hyphae 4.0 – 6.0 µm wide. Stipe made of interwoven, thin-walled generative hyphae: 3.6 – 5.0 µm wide, sclerified thick-walled generative hyphae 8.0 – 10.0 µm wide, and abundant thick-walled skeletal hyphae 4.5 – 6.0 μm wide. Generative hyphae possess clamp connections. Subhymenium composed of tightly packed thin-walled cells 3.5 – 5.0 μm broad. Basidiospores, basidia and hyphal system appear hyaline in both KOH and Melzer's reagent.

Habit, habitat and occurrence: This species is considered rare, and it appears solitary or (more often) gregariously from late spring to autumn, by growing saprotrophically or as weak parasite on logs, stumps and branches of both dead and living poplars (mainly *Populus alba* and *P. tremula*). It has a rather limited distribution in central and eastern Europe (Austria, Croatia, Czech Republic, Germany, Hungary, Slovakia, Slovenia and Ukraine) [66-69], and it is considered very rare in hemiboreal and boreal biogeographical zones of Scandinavia, recorded in Finland, Norway and Sweden [70].

Notes: *P. calyptratus* is a well distinguished species of the genus *Pleurotus*. Despite the fact that it presents several common morphological characters with *P. dryinus*, it could be discriminated by lacking a conspicuous stipe and by not producing chlamydospores, and by forming a pellicular veil. In addition, it presents higher mycelium growth rates, while production of basidiomata *in vitro* is performed faster than in the case of *P. dryinus* [62]. In addition, Hilber [71] also mentioned the occurrence of arboriform skeletal hyphae in *P. calyptratus* hyphal system, its substrate-specificity (growing only on *Populus* spp.) and that it produces its basidiomata from summer to early autumn (whereas *P. dryinus* from September to spring).

Pleurotus cornucopiae (Paulet) Rolland

Description: Pileus (2.0–) 4.0 – 15.0 (–30.0) cm, initially convex, smooth to velvet centre, circular to kidney-like, with inrolled margin, soon becoming totally smooth, slightly viscid, depressed and deeply so with age to almost umbilicate, with a straight sometimes cracked margin; when young, cream to pale yellowish brown, ivory to gray brownish buff with age. Lamellae thin, wide, crowded, deeply decurrent to the stipe base, with entire edges and conspicuous anastomoses gradually passing to anastomising ridges, pale whitish cream at first becoming ochre pale yellowish at maturity. Stipe 3.0 – 11.0 x (0.7)1 – 2(–2.5) cm; central to eccentric, rarely almost lateral, broader at top and cylindric almost equal below, slightly tapering towards the base where it is swollen again to form a foot of a group of fasciculately growing stipes; color whitish cream to ivory and somewhat darker pale grayish brown towards the base, sometimes with a lilaceous tint, surface of the upper portion completely covered by anastomising ridges, finely velvety to densely hairy at base. Flesh solid but relatively thin and fragile, whitish to pale cream. Odour pleasant sweet anise- like, somewhat farinaceous when cut. Taste mild acidic to farinaceous and occasionally spicy. Spore-print whitish to lilaceous.

Basidiospores (6.5–) 7 – 11 (–13.5) x 3.0 – 5.0 µm, subcylindric to ellipsoid, oblong, thin-walled, with a distinct and broad apiculus. Basidia 25 – 45 x 5.5 – 7.0 µm, 4-spored cylindric, thin-walled; sterigmata up to 2 µm long. Hymenial cystidia absent; instead cylindrical, elongated, clavate, rostrate or lecythiform basidioles present 20 – 35 x 5 – 7 µm in size. Caulocystidia appear as sterile hymenial elements, 16 – 20 x 3.0 – 5.5 µm. Hyphal system mono- or dimitic. Pileipellis 50 – 120 µm broad, a cutis with generative hyphae 3 – 7 µm wide. Pileus hyphal system is made of thin-walled sclerified generative hyphae, 5.5 – 12.0 µm broad, and skeletal hyphae, with no clamp connections, 4.0 – 7.0 µm broad. Lamellar hyphal system presents thin-walled generative hyphae (3–)5–13 µm wide, sclerified, thick-walled generative hyphae 1.5 – 3.0 µm wide and skeletal hyphae 3.0 – 4.0 µm wide. Stipe context is irregular and made of thin-walled generative hyphae, 6.0 – 13.0 µm wide, sclerified thick-walled generative hyphae, 8.0 – 10.0 µm wide and several skeletal hyphae 3.5 – 5.0 µm wide. Generative hyphae possess clamp connections. Subhymenium 20 – 30 µm thick composed of thin-walled cells 3.0 – 5.0 µm broad. Basidiospores, basidia and hyphal system appears hyaline in KOH and Melzer's reagent.

Habit, habitat and occurrence: It often grows in large clusters with several pilei, from late spring to autumn on various deciduous trees (genera: *Acer*, *Alnus*, *Fagus*, *Fraxinus*, *Prunus*, *Quercus* but usually on *Ulmus*). It grows as a saprotroph, but possibly also as a weak parasite on living plants. It is distributed throughout Europe although not very common [62,70–74].

Notes: *P. cornucopiae* is mainly discriminated from other taxa of the section *Pleurotus* due to the shape and color of its pileus (convex to infundibuliform, whitish to light yellow), the subcentral fistulose-dichotomic stipe, the dimitic trama and the spore-print color. Biochemical and molecular data confirmed the distinct positioning of this species within the genus *Pleurotus* [60,61,75-78], and rDNA sequences demonstrated its relative affinity with other dimitic Pleuroti such as *P. djamor* (Rumph. ex Fr.) Boedijn and *P. calyptratus* [61]. These studies also verified the close affinity with *P. cornucopiae* of morphologically-similar populations occurring in the eastern and north-eastern part of Asia under the name *P. citrinopileatus* Singer.

Pleurotus dryinus (Pers.) P. Kumm

Description: Pileus (4.0–) 7.0 – 13.0 (–18.0) cm, fleshy, at first convex, semicircular to kidney-shaped, covered with gray-brown fibrillose scales, margin inrolled and with remains of the veil, whitish, beige-buff to beige-brown, yellowing at handling becoming beige-orange; at maturity, kidney-shaped to cell-like, becoming flat, and smooth, coarse-looking due to cracking of its surface, with broad to sinuate margin. Lamellae thin, crowded, adnate to decurrent to the stipe top, with entire edges, forked or with frequent anastomoses mainly in the vicinity of the stipe, white to cream, with age, light warm yellow to light beige-orange. Stipe size 3.0 - 5.5 x 0.8 - 2.5 cm, lateral, eccentric to rarely central, cylindric to fistulose, solid, white to cream becoming slightly darker pale beige-brown when older, often with a narrow ring zone from the remains of the partial veil, longitudinally striate above and hirsute to tomentose below. The existing veil extends from the pileus margin to the upper part of the stipe, white, cream to ivory. Flesh firm and solid but elastic, white to ochraceous cream close to lamellae. Odour sometimes distinct fungoid and somewhat sweet, sometimes unpleasant. Taste mild to slightly bitter but pleasant. Spore print whitish-cream to light yellow.

Basidiospores 8.5 – 15.5 (– 17) x (2.5–) 3.0– 5(– 5.5) μm, oblong ellipsoid to cylindrical, thin-walled, possessing a small and broad apiculus. Basidia 30 – 55 (-60) x 5.0 – 10.0 μm, 2– and 4–spored, oblong clavate, thin-walled; sterigmata up to 5 μm long. Hymenial cystidia absent. Hyphal system mono- or dimitic. Pileipellis a cutis, 110 – 130 μm broad, with thin-walled generative hyphae 4 – 7 μm wide. Pileus hyphal system composed of thin-walled and sclerified thick-walled generative hyphae 5.0 – 11.5 µm wide, and skeletal hyphae (with no clamp connections) 6.2 – 9.0 µm wide. Lamellar hyphal system irregular with thin-walled generative hyphae 4.0 – 6.2 µm wide, sclerified thick-walled generative hyphae 7.0 – 11.0 (-20) µm wide, and skeletal hyphae 4.0 – 5.6 µm wide. Stipe context composed of interwoven, thin-walled generative hyphae 4.0 – 5.5 µm wide, sclerified thick-walled generative hyphae 7.0 – 11.0 µm wide, and abundant skeletal hyphae 5.0 – 6.0 µm wide. Generative hyphae possess clamp connections. Subhymenium 20 – 45 µm thick, composed of tightly packed thin-walled cells 3.0 – 4.0 µm wide. Basidiospores and basidia hyaline in KOH and Melzer's reagent; hyphal system yellowish in KOH.

Habit, habitat and occurrence: It grows gregariously or more often by forming single basidiomata, in autumn and winter, on a variety of deciduous trees, *e.g.* *Alnus*, *Betula*, *Fagus*, *Juglans*, *Malus*, *Platanus*, *Populus*, *Quercus* and *Ulmus*, and rarely on conifers, such as *Picea* and *Abies*. It grows mostly as a saprotroph, but also as weak parasite, and mostly on standing living or dead trees. This species is rather common and presents a wide distribution throughout Europe [67,68,70,73,74,79-81].

Notes: *P. dryinus* is easily distinguished from other *Pleurotus* species since it produces a distinct veil, possesses a dimitic trama, forms chlamydospores and shows very low mycelium growth rates [62,71]. *P. dryinus* is also reported to occur in north Africa, Asia, north America and New Zealand [61, 62].

Pleurotus eryngii var *eryngii* (DC.) Quél (Fig. 1)

Description: Pileus 4.0-12.5 cm, fleshy, at first convex but soon almost flat, smooth, circular to kidney- shaped with inrolled margin, dark reddish brown, or lighter warm grey brown, radially innately fibrillose or with small brown squamules; later mostly flat or concave and circular, smooth, dry, beige-buff to light beige to beige brown, with numerous beige brown innate squamules, margin even or slightly sinuate. Lamellae entire, thin, broad, moderately crowded, decurrent to stipe top, without or with few anastomoses near the stipe, cream to ivory to pale pinkish cream to light beige-orange. Stipe 1.5 – 5.0 x 0.5 – 2.5 cm, solid, almost central to eccentric, mostly cylindrical or with slightly inflated base and then with a tapering short root, whitish cream to ivory, then ochraceous brown with age, downy to almost flocculose mostly around its base. Flesh firm and often elastic, white to cream. Odour faint, mild to anise-like. Taste mild and distinctively pleasant. Spore print cream to light yellow to light brown.

Basidiospores 8.0 – 13.0 x 3.5 – 6.0 μm, ellipsoid oblong to cylindrical, thin-walled, possessing a small and broad apiculus. Basidia 30 – 52 x 7.0 – 9.0 μm, 4-spored, cylindrical to clavate, thin-walled; sterigmata up to 4 μm long. Cheilocystidia rare. Pleurocystidia absent. Caulocystidia not frequent, mostly on the upper part of the stipe. Hyphal system monomitic. Pileipellis a cutis 160 – 420 μm broad, of irregular interwoven hyphae, 5-6 μm wide, slightly thick-walled and with yellowish pigment. Pileus hyphal system formed of tightly arranged hyphae 4.0 – 5.5 μm. Lamellar hyphal system irregular, with branched hyphae 4.0 – 5.5 μm wide. Stipe made of irregular interwoven hyphae 4.5 – 8.5 μm. All hyphae possess clamp connections. Subhymenium composed of tightly packed thin-walled short and broad cells. Basidiospores, basidia and hyphal system hyaline in KOH and Melzer's reagent.

Figure 1: Basidiomata of *P. eryngii* var. *eryngii*.

Habit, habitat and occurrence: *P. eryngii* var. *eryngii* comprises fungi that are widely distributed in the northern hemisphere, growing out of the soil as a saprotroph or weak parasite on roots and residues of various plant species of the Apiaceae family. Mushrooms appear solitary or in small groups from late spring to winter. Its geographical distribution in Europe extends from the south parts of the Netherlands, Germany, Poland and Russia in the north to the entire south coastal region (Portugal, Spain, France, Italy, Greece, Turkey and Cyprus) including also most of the central European countries [8,62,71,74,79,82-88].

Notes: The *P. eryngii* species-complex includes populations of choice edible mushrooms growing in close association with different genera of the Apiaceae family. The principal discriminative features of *P. eryngii* are the white to brown colors of the pilei which are scattered with several squamules, the robust and solid central to subcentral stipe, the relatively large size of basidiospores and the habitat [15,62]. Previous studies on this group employing morphological and compatibility data arrived at the conclusion that this species could be separated into three host-associated varieties, namely var. *eryngii*, var. *ferulae* and var. *nebrodensis* [62,71]. However, the application of combined approaches using morphological characters and ecological features with isozyme, PCR-RAPD and DNA sequencing provided answers to this perplexed situation [15,89,92-93]. Results confirmed classification of *Pleurotus* strains into "eco-groups" mainly in accordance to their host-specificity: *Pleurotus* isolates growing on *Cachrys ferulacea* presented increased genetic distances with all other populations, representing thus a distinct species, *P. nebrodensis* (Inzenga) Quél. Most of the other individuals studied were grouped within a larger cluster (*P. eryngii* sensu stricto) subdivided into groups largely corresponding to other hosts, and thus constituting taxa at the varietal level: *P. eryngii* vars. *elaeoselini*, *eryngii*, *ferulae*, and *thapsiae*. All Mediterranean *Pleurotus* populations growing on Umbellifers seem to have recently diverged through a sympatric speciation process (*i.e.* speciation without geographical isolation: establishment of isolation within a community of potentially interbreeding individuals which are morphologically similar/identical), that is based on both intrinsic reproductive barriers and extrinsic ecogeographical factors [15,92].

Pleurotus eryngii var. *elaeoselini* Venturella, Zervakis & La Rocca

Description: Pileus 5 – 14 (–20) cm, fleshy, at first whitish, with a velvety and opaque surface, convex, with inrolled margin; at maturity flat to finally depressed, seldom almost funnel-shaped, smooth, bright and greasy, with alutaceous tones or flesh-colored; often lacerated in small appressed areolae, evident on the disk and contrasting with the white-ivory colored flesh in the background; margin thin, acute, flat and irregular, sometimes also lobate. Lamellae thick, arcuate, mixed by many lamellulae, deeply decurrent, entirely white, margin entire, concolorous, lightly flesh-colored in the ripe basidiomata; the unripe basidiomata shows evident anastomoses on the stipe, in some cases disappearing with the growth. Stipe 4 – 7.5 x 1.2 – 2.8 (–3.6) cm, sturdy, filled and firm, irregularly cylindrical, attenuate at the base, sometimes radicating, central to eccentric, concolorous with the cap, lightly pruinose in the unripe basidiomata, then smooth, glabrous and yellowish at the base. Flesh firm and compact, fibrous, elastic, white, with a fungus-like smell and taste.

Basidiospores 10.0 – 14.0 x 5.0 – 7.0 μm, cylindrical to irregularly ellipsoid, smooth, hyaline, guttulate with pronounced apiculus. Basidia, clavate, 4-spored, 30 – 50 x 8 – 12 (–14) μm, sterigmata 4-6 μm. True hymenial cystidia absent but with clavate cystidia-like elements, basidioles thinning at apex, sometimes showing a thin acute beak 40 – 65 x 8 – 12 μm in size. Pileipellis cutis-like, with closely appressed cylindrical hyphae 5 – 12 μm wide.

Habit, habitat and occurrence: *P. eryngii* var. *elaeoselini* appears mostly in spring but it has been also collected in autumn up to early November, found in the same habitats with *P. eryngii* var. *ferulae*, occurring from sea- level to 1200 m. *P. eryngii* var. *elaeoselini* was initially described by Venturella *et al.* [89] from Sicily (Italy) on *Elaeoselinum asclepium* (L.) Bertol. ssp. *asclepium*. Later, it has been also reported to occur in Spain in association with *Thapsia villosa* L., *Ferula communis* L., *Magydaris panacifolia* (Vahl) Lange and *Elaeoselinum gummiferum* (Desf.) Samp. [90], and in Romania on *Laserpitium latifolium* L. roots [94].

Pleurotus eryngii var. *ferulae* (Lanzi) Sacc. (Fig. 2)

Description: Pileus 5 – 25 (–30) cm, fleshy, at first light brown to beige brown, convex, with inrolled margin; at maturity flat to finally depressed, smooth to suede like, dark brown to chestnut brown to buff brown to grey-brown to slate grey, often with darker colored squamules or innatae fibrillae; margin thin, acute, flat and irregular, sometimes also lobate. Lamellae thick, mixed by many lamellulae, deeply decurrent, white to cream white to light beige, often anastomizing by the stipe, margin entire, concolorous, lightly flesh-colored in the ripe basidiomata; the unripe basidiomata show evident anastomoses on the stipe, in some cases disappearing with the growth. Stipe 3 – 10 x 1 – 4 cm, sturdy, filled and firm, irregularly cylindrical, attenuate at the base, sometimes radicating, central to eccentric, concolorous with the cap, lightly pruinose in the unripe basidiomata, then smooth, glabrous and beige-brown towards the base. Flesh firm and compact, thick, velvety, pruinose, white often heavily pigmented, with terminal club-like cells, and a fungus-like smell and taste.

Basidiospores 9.2 – 13.9 x 4.5 – 7.2 μm, cylindrical to irregularly ellipsoid, smooth, hyaline, guttulate with pronounced apiculus. Basidia, clavate, 4-spored, 31 – 48 x 6.5 – 10 μm, sterigmata 4 – 6 μm. True hymenial cystidia absent but cystidia-like clavate, basidioles present thinning at apex.

Figure 2: Basidiomata of *P. eryngii* var. *ferulae.*

Habit, habitat and occurrence: *P. eryngii* var. *ferulae* appears from autumn to spring, singly or in clusters. It is found in garigues, wastelands and pastures, on limestone and silicaceous soils, occurring from sea-level to 1300 m. It is widespread in Europe throughout the zone of occurrence of its associated plant *Ferula communis.*

Pleurotus eryngii var. *thapsiae* Venturella, Zervakis & Saitta

Description: *P. eryngii* var. *thapsiae* differs from vars. *eryngii* and *ferulae* in producing significantly smaller and darker basidiomata with pileus 2 – 10 cm in size and dark brown colours. The basidiospores in *P. eryngii* var. *thapsiae* are irregularly ellipsoid to cylindrical (10 – 14 x 5 – 7 μm, Q = 1.9 – 2.1). This variety possesses also cheilocystidia-like basidioles with an attenuate apex at the top, and sometimes with an acute beak, showing a close resemblance to *P. eryngii* var. *elaeoselini* microscopic features. In addition, the *P. eryngii* var. *eryngii* basidiomata are produced singly or in clusters from early autumn to winter, while basidiomata of *P. eryngii* var. *thapsiae* appear mostly singly and in spring. Moreover, the former is quite common and occurs from sea level to 1500 m, while the latter is infrequent and restricted to the mountainous zone.

Habit, habitat and occurrence. *Pleurotus eryngii* var. *thapsiae* is a spring mushroom and the appearance of the basidiomata is restricted from the end of March until May, growing mostly singly on roots and stems of *Thapsia garganica*. It is only known to occur in the Madonie Mts. (north Sicily) collected on calcareous soils, in arid pastures and at elevations ranging from 1000 to 1500 m [91].

Pleurotus fuscosquamulosus Reid & Eicker (*P. cystidiosus* O.K. Miller sensu lato) (Fig. 3)

Description: Pileus (3.0–) 8.5 – 14.0 (–18.0) x (2.6–) 4.5 – 6.3 (– 7.5) cm, semicircular to circular, at first smooth, flat or slightly convex, with even and inrolled margins, brown to beige-brown to beige-orange; at maturity flat to concave, light beige to orange-beige to light warm yellow at the periphery and beige-buff to beige-orange to light-brown to darker brown towards the centre, surface sometimes covered with orange-beige squamules. Lamellae thin, wide, relatively dense, decurrent to the stipe top, with even edges and rare anastomoses mainly near the stipe, white to light warm yellow to light beige-orange and presence of long lamellulae. Stipe (0.5–) 1.5 x 4.0 (–5.5) cm, eccentric, thick, smooth, cylindric to fistulose, cream to beige-buff, often with coremia in its lower part. Absence of any kind of veil. Spore print colour (dry) light warm yellow to orange-beige to yellow-orange.

Basidiospores 10.5 – 18.5 x 3.0 – 6.5 μm, cylindric to oblong, thin walled, inamyloid. Absence of any kind of chlamydospores. Basidia 32 x 9.5 μm at the top and 4.0 – 5.0 μm at the basis, cylindric, bearing four sterigmata 5 μm long. Cheilocystidia 26 – 36 x 6.8 – 9.8 μm, frequent, pyriform, thin-walled. Pileocystidia 21 – 37.9 x 5.7 – 8.8 μm, subcylindric to cylindric, rather thick walled. Cuticle of pileus 40 – 57 μm and trama monomitic comprised of generative, thin-walled hyphae (4.3 – 8.5 μm) with frequent clamp-connections. Trama of lamellae 2.7 – 5.8 μm and stipe 5.0 – 8.4 μm, irregular, with thin-walled generative hyphae. All hyphae and cystidia inamyloid.

Synnematoid fructifications present, on the surface of lamellae and stipe, and on mycelial cultures. These fructifications are composed of white coremia with a black mass of allantoid conidia on their top (av. size of conidia: 10.2 – 22.0 x 4.1 – 6.4 μm). The former have the structure of a synnematal column 2 – 4 mm high, formed by tightly compacted masses of hyphae which at one side bear the conidia in chains in a basipetal succession, while on the other, end up by building a dense pseudoparenchymatous tissue around the conidiophore's base. The latter are thick-walled (double-layered light brown walls) appearing with remnants on their surface of the mucus which surrounds them.

Figure 3: *P. fuscosquamulosus* on a decayed trunk of *Ficus carica* L.

Habit, habitat and occurrence: This species is very rare in Europe reported until now only from Greece (in Attika region, and in Salamina and Lesvos islands) as *P. cystidiosus* [65,76]. It usually forms single mushrooms on dead standing trees of *Ficus carica* and *Populus* spp., from autumn to spring.

Notes: The subgenus *Coremiopleurotus* Hilber is composed of *Pleurotus* taxa producing asexual synnematoid anamorphs (the imperfect state belongs to the genus *Antromycopsis* Pat. & Trab.) presenting a world-wide distribution [65,95]. Members of this group also possess several distinct taxonomic characteristics, incl. the basidiospores shape (oblong to elliptical), the formation of abundant pileocystidia and cheilocystidia, the very low mycelium growth rate *etc.* [62,65]. *P. cystidiosus* sensu lato (*i.e.* incl. *P. fuscosquamulosus*) encompasses populations in USA, Africa and Europe which are partially or (rarely) fully intercompatible [65]; *Pleurotus smithii* is restricted to Latin America and it is morphologically similar to *P. cystidiosus*, and *P. australis* is found in Australia and New Zealand demonstrating distinct anatomical and physiological characters (both for the teleomorph and the anamorph stage). In addition, *P. abalonus* comprises isolates of Asiatic origin with a recognizable phenotype and a limited ability to interbreed with other synnematoid *Pleurotus* isolates of distant geographic origin [65]. Recent information based on ITS rDNA sequencing of synnematoid *Pleurotus* taxa with a world-wide origin revealed that the *P. cystidiosus* species-complex is paraphyletic [63]: it is divided into three major lineages representative of an allopatric mode of evolution: New World (incl. Mexico), Europe/Africa, and Asia/Pasific; the two latter seem to have diverged more recently. Phylogeographic analysis and evaluation of mating (high compatibility percentages within these groups in conjunction with the low intercontinental successful matings) and morphology (minute differences, generally restricted to limited variation of pileus color and cystidia anatomy) data suggest that speciation in this fungal group is associated with isolation of populations in geographically confined areas [63]. Another species thought to be related to *P. cystidiosus* is *P. gemmellarii* (Inzeng.) Sacc., isolated from Sicily in 1865. Since then, only Pegler [98] reported the occurrence of a similar looking fungus in Pakistan referring to it as *Pleurotus* aff. *gemmellarii*. Later, Nair and Kaul [99], described *Antromycopsis broussonetiae* var. *minor* from India and considered it to be the anamorph of Pegler's isolate.

However, their fungus should belong either to *P. cystidiosus* or to *P. smithii* [96,97]. In addition, the European coremioid *Pleurotus* already studied apart from occurring along with its imperfect state, differs from *P. gemmellarii* in the size and colour of pileus and stipe, and in not producing any pigments in the subhymenium [12].

Pleurotus nebrodensis (Inzenga) Quél

A detailed description and comments for this species are provided in the next chapters.

Pleurotus ostreatus (Jacq.) P. Kumm. (Fig. 4)

Description: Pileus 4.0-16.0 (-33) cm at first convex soon flattened, semicircular to kidney-shaped or somewhat rounded triangular with inrolled margin, smooth to velvety, subviscid, dark brownish grey to metal to steel, often bluish grey or sometimes brown to beige-orange to orange-brown; at maturity, flat to slightly convex, spathulate to oyster to kidney-shaped, smooth, or slightly pubescent at centre, moist light to warm brown fading on drying to beige-buff to brown or light blue-grey to light warm grey to light grey to warm grey to nickel to grey to platinum sometimes with darker appressed radial squamules, with even to slightly inrolled, and undulating margin, sulcate at times. Lamellae thin, broad, crowded, decurrent, entire, with anastomoses, or forked towards the stipe apex, with abundant lamellulae, white to light beige, or ivory to light warm grey. Stipe sometimes also absent and mostly short, rarely long, 0.5–10 x 1.5 – 5.5 cm; lateral to eccentric, solid, cylindrical and connate, smooth to longitudinally striate, villous or with floccules to strigose near its base, white to cream to ivory. Flesh firm, often thin, elastic and occasionally fragile, white to cream to ochraceous, especially under pileipellis and close to lamellae. Odour faint, fungoid, usually neutral. Taste mild, pleasant, fungoid to astringent. Spore-print whitish to cream to ivory to pale lilaceous grey.

Basidiospores (6.5–) 8.0 – 12.5(–13.5) x (2.0)3.0 – 4.59 (5.5) μm, subcylindrical to cylindrical to bacilliform, with suprahilar depression and a small apiculus, thin-walled, hyaline in KOH and Melzer's reagent. Basidia 20 – 45 x 3.0 – 8.5 μm, cylindrical to clavate, hyaline, thin-walled; 4–spored, a few 2–spored; basidioles often rostrate, sterigmata up to 4 μm long. Hymenial cystidia absent but some lecythiform cystidia-like basidioles may exist, and then 20 – 30 x 3 – 7 μm. Hyphal system monomitic. Pileipellis a compact cutis of 100 - 180 μm broad, formed of rather thin-walled generative hyphae. Pileus hyphal system formed of rather thick-walled and oleiferous hyphae 5.0 – 13.0 μm. Lamellar hyphal system irregular, with interwoven oleiferous, rather thick-walled hyphae 4.0 – 9.0 μm wide. Stipe composed of thin-walled and of sclerified thick-walled hyphae 5.0 – 11.0 μm in diam. Subhymenium up to 30 μm broad, composed of short thin-walled hyphae 2.5 – 4.5 μm. All generative hyphae with clamp connections.

Figure 4: Basidiomata of *P. ostreatus* (Jacq.) P. Kumm. on decayed trunk.

Habit, habitat and occurrence: It usually appears in large clusters, seldom solitary, from early autumn to winter; it grows on trunks of many different deciduous and coniferous trees (*Abies, Alnus, Betula, Castanea, Fagus, Juglans, Picea, Populus, Quercus, Salix, Ulmus, Castanea etc.*). It is fairly common and it is widely distributed throughout Europe [62,72,74,100].

Notes: *P. ostreatus* is the type species of the genus *Pleurotus*; it is distinguished by the dark colored (brown to gray) pilei, the dense lamellae with frequent anastomoses, the lateral to eccentric stipe, the monomitic trama and the ability to produce basidiomata at low temperatures on a wide range of substrates. In the past, several other putative taxa were associated with *P. ostreatus*, *e.g.* *P. columbinus* Quél. (a European taxon) was at first considered as a distinct species because of its blue-gray pilei, the coniferous host and the alleged mating incompatibility with *P. ostreatus* [72,101]. However, subsequent studies established that *P. columbinus* is a variety of *P. ostreatus* [62,71,102] *i.e.* *P. ostreatus* var. *columbinus* (Quél.) Quél., despite the fact that it presented partial compatibility with several *P. ostreatus* strains and a close yet distinct positioning in dendrograms resulting from biochemical and molecular analyses [75,76,77]. Similar issues arose in the past with the use of other names either for similar looking taxa (*e.g.* *P. salignus* (Pers.) P. Kumm.) or for specimens whose identity was not thoroughly assessed (*e.g.* *P. floridanus* Singer or P. "florida"); in most cases, they proved to be infraspecific taxa under *P. ostreatus*.

Pleurotus pulmonarius (Fr.) Quél

Description: Pileus (1.5 –) 3 – 11 (– 12) cm, at first convex but soon flat, leather-like spathulate to kidney-shaped, light brown to beige brown to beige-buff to orange brown and margin inrolled; at maturity, circular to spathulate to lung-shaped, usually flat but sometimes concave towards the centre or slightly infundibular, pale yellowish-brown to greyish brown to light beige to almost whitish, smooth, shiny, or slightly arachnoid, rarely somewhat squamose, more frequently tomentose close to the attachment point with stipe. Lamellae thin, crowded, decurrent to the stipe top, often intervenose along its entire length, with rare anastomoses, white to cream to ivory, with entire or somewhat serrulate, concolorous or slightly darker brown edge. Stipe absent or short 1.5 – 5 x 2 – 4 cm; eccentric to lateral cylindrical to connate sometimes, solid, white to cream, smooth to slightly villous to tomentose near its base. Flesh firm and elastic, white to cream. Odour mild, fungoid to distinctly sweetish or anise-like. Taste distinct, fungoid and pleasant to somewhat bitterish. Spore-print cream to light beige to beige-buff.

Basidiospores 7.5 – 14.5 x 2.5 – 5.0 µm, subcylindrical to cylindrical to bacilliform, thin-walled, with suprahilar depression and a small apiculus. Basidia

20 – 54 x 3.5 – 8.5 µm, cylindric to clavate, thin- walled; 4-spored, a few 2-spored, sterigmata up to 4 µm long. Hymenial cystidia absent, but some lecythiform cystidia-like basidioles may exist, and then up to 50 x 2 – 4 µm, with branched or unbranched projections. Hyphal system monomitic. Pileipellis a compact cutis up to 70 µm broad, with thin-walled hyaline generative hyphae. Pileus hyphal system formed of irregular, thin- walled and of sclerified thick-walled hyphae 2.5 – 15.0 µm wide. Lamellar hyphal system irregular with interwoven thin-walled and with sclerified thick-walled hyphae, 4.0 – 12.0 µm. Subhymenium around 20 µm broad, with short and thin-walled hyphae, 3.5-6.0 µm. Stipe composed of thin-walled and sclerified thick-walled hyphae, 3.4 – 9.6 µm in diam. Generative hyphae possess clamp connections. Basidiospores, basidia and hyphal system appearing hyaline in KOH and Melzer's reagent.

Habit, habitat and occurrence: It appears from late spring to autumn in north Europe and until early winter in the southernmost regions of the continent, growing solitary or usually in clusters and groups, saprotrophic or probably also weakly parasitic on trunks of angiosperms of the genera *Aesculus, Betula, Fagus, Fraxinus, Populus, Sorbus, Quercus etc.* It grows throughout Europe [60,62].

Notes: In the past, there were considerable ambiguities on the taxonomic relationship between *P. pulmonarius* and *P. ostreatus* because of their significant morphological similarity and overlapping occurrence [101,102]. For elucidating this situation, Hilber [71], Zervakis and Balis [62], Vilgalys *et al.* [103] and Petersen and Hughes [60] demonstrated by mating experiments that *P. pulmonarius* and *P. ostreatus* are two distinct biological species. Additional evidence was provided by isozyme, RFLP, rDNA sequencing and FT-IR spectroscopy approaches [61,75-78]. In the past, the use of the name 'P. sajor-caju' produced confusions since it was considered as a distinct *Pleurotus* species (unfortunately, the use of strains named as 'P. sajor-caju' continues until now both among scientists and within the commercial mushroom sector). However, it is clear from the results of both compatibility and DNA analyses that such strains belong to *P. pulmonarius* [62,71,75,77]. Similarly, many strains originating from North America were identified as *P. sapidus*, but they turned out to be either *P. pulmonarius* or *P. ostreatus* [62,78,103].

ACKNOWLEDGEMENTS

Apart from the authors (G.I. Zervakis and E. Polemis), G. Koutrotsios and G. Kallontzis also provided *Pleurotus* photos from their own personal archive for use in this chapter. This material is copyrighted and remains property of the photographers.

CONFLICT OF INTEREST

The author(s) confirm that this chapter content has no conflict of interest.

REFERENCES

[1]   Jeffries TW. Biodegradation of lignin-carbohydrate complexes. Biodegradation 1990; 1: 163-176.

[2]   Kirk TK, Farell RL. Enzymatic 'combustion': the microbial degradation of lignin. Ann Rev Microbiol 1987; 41: 465-505.

[3]   Mayson E, Verachtert H. Growth of higher fungi on wheat straw and their impact on the digestibility of the substrate. Appl Microbiol Biot 1991; 36: 421-424.

[4]   Chang S.T. Dervelopment of the culinary-medicinal mushrooms industry in China: past, present and future. Int J Med Mushrooms 2006; 8: 1-17.

[5]   Ragunathan R, Gurusamy R, Palaniswamy M, Swaminathan K. Cultivation of *Pleurotus* spp. on various agro-residues. Food Chem 1996; 55: 139-144.

[6]   Royse DJ, Rhodes TW, Ohga S, Sanchez JE. Yield, mushroom size and time to production of *Pleurotus cornucopiae* (oyster mushroom) grown on switch grass substrate spawned and supplemented at various rates. Bioresource Technol 2004; 91: 85-91.

[7]   Salmones D, Mata G, Waliszewski KN. Comparative culturing of *Pleurotus* spp. on coffee pulp and wheat straw: Biomass production and substrate biodegradation. Bioresource Technol 2005; 96: 537-544.

[8]   Rodriguez Estrada AE, Royse DJ. Yield, size and bacterial blotch resistance of *Pleurotus eryngii* grown on cottonseed hulls/oak sawdust supplemented with manganese, copper and whole ground soybean. Bioresource Technol 2007; 98: 1898-1906.

[9]   Kalmis E, Azbar N, Yıldız H, Kalyoncu F. Feasibility of using olive mill effluent (OME) as a wetting agent during the cultivation of oyster mushroom, *Pleurotus ostreatus*, on wheat straw. Bioresource Technol 2008; 99: 164-169.

[10]   Ruiz-Rodriguez A, Soler-Rivas C, Polonia I, Wichers HJ. Effect of olive mill waste (OMW) supplementation to oyster mushrooms substrates on the cultivation parameters and fruiting bodies quality. Int Biodeter Biodegr 2010; 64: 638-645.

[11]   Zervakis G. Cultivation of the king-oyster mushroom *Pleurotus eryngii* (DC.:Fr.) Quél. on substrates deriving from the olive-oil industry. Int J Med Mushrooms 2005; 7: 486-487.

[12]   Zervakis G, Balis C. Comparative study on the cultural characters of *Pleurotus* species under the influence of different substrates and fruiting temperatures. Micol Neotrop Apl 1992; 5: 39-47.

[13] Zervakis G, Venturella G. Mushroom breeding and cultivation favors *ex situ* conservation of Mediterranean *Pleurotus taxa*. In: Engels JMM, Ramanantha Rao V, Brown AHD, Jackson MT, Eds. Managing Plant Genetic Diversity. IPGRI, CABI Publishing 2002; pp. 351-358.

[14] Zervakis G, Yiatras P, Balis C. Edible mushrooms from olive mill wastes. Int Biodeter Biodegr 1996; 38: 237-243.

[15] Zervakis G, Philippoussis A, Ioannidou S, Diamantopoulou P. Mycelium growth kinetics and optimal temperature conditions for the cultivation of edible mushroom species on lignocellulosic substrates. Folia Microbiol 2001; 46: 231-234.

[16] Zervakis G, Venturella G, Papadopoulou K. Genetic polymorphism and taxonomic relationships of the *Pleurotus eryngii* species-complex as resolved through the analysis of random amplified DNA patterns, isozyme profiles and ecomorphological characters. Microbiology 2001; 147: 3183-3194.

[17] Gunde-Cimerman N. Medicinal value of the genus *Pleurotus* (Fr.) P. Karst. (*Agaricales* s.l. *Basidiomycetes*). Int J Med Mushrooms 1999; 1: 69-80.

[18] Wasser SP. Medicinal mushrooms as a source of antitumor and immunomodulating polysaccharides. Appl Microbiol Biot 2002; 60: 258-274.

[19] Wasser SP, Weis AL. Theurapetic effects of substances occurring in higher *Basidiomycetes* mushrooms: a modern perspective. Crit Rev Immunol 1999; 19: 65-96.

[20] Babitskatya UG, Sherba VV, Mitropolskaya NY, Bisko NA. Exopolysaccharides of some medicinal mushrooms: production and composition. Int J Med Mushrooms 2000; 2: 51-54.

[21] Gunde-Cimerman N, Plemenitas A. Hypocholesterolemic activity of the genus *Pleurotus* (Fr.) Karst. (*Agaricales* s.l., *Basidiomycetes*). Int J Med Mushrooms 2001; 3: 395-397.

[22] Alarcón J, Águila S, Arancibia-Avila P, Fuentes O, Zamorano-Ponce E, Hernández M. Production and purification of statins from *P. ostreatus* (*Basidiomycetes*) strains. Z Naturforsch 2003; 58: 62-64.

[23] Zhang M, Zhang L, Keung Cheung PC, Choon Ooi VE. Molecular weight and anti-tumor activity of the water-soluble polysaccharides isolated by hot water and ultrasonic treatment from the sclerotia and mycelia of *Pleurotus tuber-regium*. Carbohyd Polym 2004; 56: 123-128.

[24] Gregori A, Svagelj M, Pohleven J. Cultivation techniques and medicinal properties of *Pleurotus* species. Food Technol Biotech 2007; 45: 238-249.

[25] Synytsya A, Míčková K, Synytsya A, Jablonskỳ I, Spěváček J, Erban V, Kováříková E, Čopíková J. Glucans from fruit bodies of cultivated mushrooms *P. ostreatus* and *Pleurotus eryngii*: structure and potential prebiotic activity. Carbohyd Polym 2009; 76: 548-556.

[26] Tsai SY, Huang SJ, Lo SH, Wu TP, Lian PY, Mau JL. Flavor components and antioxidant properties of several cultivated mushrooms. Food Chem 2009; 113: 578-584.

[27] Tripothi JP, Yadar JS. Optimization of solid substrate fermentation of wheat straw into animal feed by *Pleurotus ostreatus* - a pilot effort. Anim Feed Sci Tech 1992; 37: 59-72.

[28] Hadar Y, Kerem Z, Gorodecki B, Ardon O. Utilization of lignocellulosic waste by the edible mushroom, *Pleurotus*. Biodegradation 1992; 3: 189-205

[29] Jalc D, Nerud F, Zitnan R, Siroka P. The effect of white-rot basidiomycetes on chemical composition and *in vitro* digestibility of wheat straw. Folia Microbiol 1996; 41: 73-75.

[30] Adamovic M, Grubic G, Milenkovic I, Jovanovic R, Protic R, Sretenovic L, Sticevic L. The biodegradation of wheat straw by *Pleurotus osteratus* mushrooms and its use in cattle feeding. Anim Feed Sci Tech 1998; 71: 357-362.

[31]   Li X, Pang Y, Zhang R. Compositional changes of cottonseed hull substrate during *P. ostreatus* growth and the effects on the feeding value of the spent substrate. Bioresource Technol 2001: 80: 157-161.

[32]   Bae JS, Kim YI, Jung SH,. Oh YG, Kwak WS. Evaluation on feed-nutritional value of spent mushroom (*Pleurotus ostereatus*, *Pleurotus eryngii*, *Flammulina velutipes*) substrates as a roughage source of ruminants. Kor J Anim Sci Technol 2006; 48: 237-246.

[33]   Hatakka A. Lignin-modifying enzymes from selected white-rot fungi: production and role in lignin degradation. FEMS Microbiol Rev 1994; 13: 125-135.

[34]   Cohen R, Persky L, Hadar Y. Biotechnological applications and potential of woo-degrading mushrooms of the genus *Pleurotus*. Appl Microbiol Biot 2012; 58: 582-594.

[35]   Bezalel L, Hadar Y, Fu PL, Freeman JP, Cerniglia CE. Metabolism of phenanthrene by the white rot fungus *Pleurotus ostreatus*. Appl Environ Microbiol 1996; 62: 2547-2553.

[36]   Haritash AK, Kaushik CP. Biodegradation aspects of Polycyclic Aromatic Hydrocarbons (PAHs): A review. J Hazard Mater 2009; 169: 1-15.

[37]   Baldrian P, Wiesche C, Gabriel J, Nerud F, Zadrazil F. Influence of cadmium and mercury on activities of ligninolytic enzymes and degradation of polycyclic aromatic hydrocarbons by *Pleurotus ostreatus* in soil. Appl Environ Microbiol 2000; 66: 2471-2478.

[38]   Byss M, Elhottová D, Třiska J, Baldrian P. Fungal bioremediation of the creosote-contaminated soil: influence of *Pleurotus ostreatus* and *Irpex lacteus* on polycyclic aromatic hydrocarbons removal and soil microbial community composition in the laboratory-scale study. Chemosphere 2008; 73(9): 1518-1523.

[39]   Pointing SB. Feasibility of bioremediation by white-rot fungi. Appl Microbiol Biot 2001; 57: 20-33.

[40]   Purnomo AS, Mori T, Kamei I, Nishii T, Kondo R. Application of mushroom waste medium from *Pleurotus ostreatus* for bioremediation of DDT-contaminated soil. Int Biodeter Biodegr 2010; 64(5): 397-402.

[41]   McMullan G, Meehan C, Conneely A, Kirby N, Robinson T, Nigam P, Banat IM, Marchant R, Smyth WF. Microbial decolourization and degradation of textile dyes. Appl Microbiol Biot 2001; 56: 81-87.

[42]   Eichlerová I, Homolka L, Nerud F. Ability of industrial dyes decolorization and ligninolytic enzymes production by different *Pleurotus* species with special attention on *Pleurotus calyptratus*, strain CCBAS 461. Process Biochem 2006; 41: 941-946.

[43]   Aggelis G, Ehaliotis C, Nerud F, Stoychev I, Lyberatos G, Zervakis G.I. Evaluation of white-rot fungi for detoxification and decolorization of effluents from the green olives debiterring process. Appl Microbiol Biot 2002; 59: 353-360.

[44]   Ntougias S, Baldrian P, Ehaliotis C, Nerud F, Merhautová V, Zervakis GI. Biodegradation and detoxification of olive mill wastewater by selected strains of the mushroom genera *Ganoderma* and *Pleurotus*. Chemosphere 2012; 88: 620-626.

[45]   Amitai G, Adani R, Sod Moriah G, Rabinovitz I, Vincze A, Leader H, Chefetz B, Leibovitz-Persky L, Friesem D, Hadar Y. Oxidative biodegradation of phosphorothiolates by fungal laccase. FEBS Lett 1998; 438: 195-200.

[46]   Singh BK, Walker A. Microbial degradation of organophosphorus compounds. FEMS Microbiol Rev 2006; 30: 428-471.

[47]   Thorn RG, Barron GL. Carnivorous mushrooms. Science 1984; 224: 76-78.

[48]   Barron GL, Thorn RG. Destruction of nematodes by species of *Pleurotus*. Can J Bot 1987; 64: 774-778.

[49] Hibbett DS, Thorn RG. Nematode trapping in *Pleurotus tuberregium*. Mycologia 1994; 86: 696-699.

[50] Redhead SA, Ginns JH. A reappraisal of agaric genera associated with brown rots of wood. Trans Mycol Soc Japan 1985; 26: 349-381.

[51] Corner EJH. The agaric genera *Lentinus*, *Panus*, and *Pleurotus* with particular reference to Malaysian species. Nova Hedwigia 1981, 69: 1-169.

[52] Kühner R. Les Hyménomycètes agaricoïdes: étude générale et classification. Lyon: Société Linnéenne de Lyon 1980.

[53] Pegler DN. Agaric Flora of the Lesser Antilles. London: Kew Bull Additional Series 1983.

[54] Singer R. The *Agaricales* in Modern Taxonomy. 4th ed. Koenigstein: Koeltz Scientific Books 1986.

[55] Hibbett DS, Vilgalys R. Phylogenetic relationships of *Lentinus* (*Basidiomycotina*) inferred from molecular and morphological characters. Syst Bot 1993; 18: 409-433.

[56] Hibbett DS, Pine EM, Langer E, Langer G, Donoghue MJ. Evolution of gilled mushrooms and puffballs inferred from ribosomal DNA sequences. P Natl Acad Sci USA 1997; 94: 12002-12006.

[57] Moncalvo JM, Lutzoni F, Rehner SA, Johnson J, Vilgalys R. Phylogenetic relationships of agaric fungi based on nuclear large subunit ribosomal DNA sequences. Syst Biol 2000; 49: 278-305.

[58] Thorn RG, Moncalvo J-M, Reddy CA, Vilgalys R. Phylogenetic analyses and the distribution of nematophagy support a monophyletic *Pleurotaceae* within the polyphyletic pleurotoid-lentinoid fungi. Mycologia 2000; 92: 241-252.

[59] Kwok OCH, Plattner R, Weisleder D, Wicklow DT. A nematicidal toxin from *Pleurotus ostreatus* NRRL 3526. J Chem Ecol 1992; 18: 127-136.

[60] Petersen RH, Hughes KW. Intercontinental interbreeding collections of *Pleurotus pulmonarius*, with notes on other species. Sydowia 1993; 45: 139-152.

[61] Vilgalys R, Sun BL. Ancient and recent patterns of geographic speciation in the oyster mushroom *Pleurotus* revealed by phylogenetic analysis of ribosomal DNA sequences. P Natl Acad Sci USA 1994; 91: 4599-4603.

[62] Zervakis G, Balis C. A pluralistic approach on the study of *Pleurotus* species, with emphasis on compatibility and physiology of the European morphotaxa. Mycol Res 1996; 100: 717-731.

[63] Zervakis GI, Moncalvo J-M, Vilgalys R. Molecular phylogeny, biogeography and speciation of the mushroom species *Pleurotus cystidiosus* and allied taxa. Microbiology-SGM 2004; 150: 715-726.

[64] Petersen RH, Ridley GS. A New Zealand *Pleurotus* with multiple-species sexual compatibility. Mycologia 1996; 88: 198-207.

[65] Zervakis G. Mating competence and biological species within the subgenus *Coremiopleurotus*. Mycologia 1998; 90: 1063-1074.

[66] Pilát A. *Pleurotus* Fries. In: Atlas des Champignons de I'Europe. Band 2. Prague 1935.

[67] Tortic M, Hocevar S. Some lignicolous *Macromycetes* from Krakovski Gozd, new or rare in Yugoslavia. Acta Bot Croat 1977; 36: 145-152.

[68] Hilber O. Biosystematische Untersuchungen zur Kenntnis von *Pleurotus calyptratus* (Lindbl. in Fr.) Sacc. und *Pleurotus dryinus* (Pers. ex Fr.) Kummer. Zeitung der Mykologie 1981; 47: 2742.

[69]   Prylutsky OV. Distribution, ecological features and conservation of *Pleurotus calyptratus* (*Agaricales*) in Ukraine. Ukr Bot J 2011; 68(5): 780-784.

[70]   Knudsen H, Vesterholt J, Eds. Funga Nordica. Nordsvamp, Copenhagen 2008; p. 965.

[71]   Hilber O. Die Gattung *Pleurotus* (Fr.) Kummer unter besonderer Berücksichtigung des *Pleurotus eryngii*-Formenkomplexes. Bibliotheca Mycologica 87. Vaduz: J. Cramer 1982.

[72]   Romagnesi H. Sur les *Pleurotus* du groupe ostreatus (*Ostreomyces* Pilát). Bull Soc Mycol Fr 1969; 85: 305-314.

[73]   Watling R, Gregory N.M. British Fungus Flora: Agarics and Boleti. Vol 6. *Crepidotaceae* and other pleurotoid agarics. Royal Botanic Garden: Edinburgh, Scotland. 157 p. 1989.

[74]   Bas C, Kuyper ThW, Noordeloos ME, Vellinga EC, Eds. Flora Agaricina Neerlandica - Critical monographs on the families of agarics and boleti occurring in the Netherlands. Volume 2: *Pluteaceae, Tricholomataceae*. Rotterdam: Balkema AA 1990; p. 137.

[75]   Iraçabal B, Zervakis G, Labarère J. Molecular systematics of the genus *Pleurotus*: analysis of restriction polymorphisms in ribosomal DNA. Microbiology 1995; 141: 1479-1490.

[76]   Zervakis G, Labarère J. Taxonomic relationships within the fungal genus *Pleurotus* as determined by isoelectric focusing analysis of enzyme patterns. J Gen Microbiol 1992; 138: 635-645.

[77]   Zervakis G, Sourdis J, Balis C. Genetic variability and systematics of eleven *Pleurotus* species based on isozyme analysis. Mycol Res 1994; 98: 329-341.

[78]   Zervakis GI, Bekiaris G, Tarantilis PA, Pappas CS. Rapid strain classification and *taxa* delimitation within the edible mushroom genus *Pleurotus* through the use of diffuse reflectance infrared Fourier transform (DRIFT) spectroscopy. Fungal Biol 2012; 116: 715-728.

[79]   Venturella G. A check-list of Sicilian fungi. Bocconea 1991; 2: 1-221.

[80]   Dimou D, Zervakis GI, Polemis E. Mycodiversity studies in selected ecosystems of Greece: I. Macrofungi from the southernmost *Fagus* forest in the Balkans (Oxya Mountain, central Greece). Mycotaxon 2002; 82: 177-205.

[81]   Dimou DM, Zervakis GI, Polemis E. Mycodiversity studies in selected ecosystems of Greece: 4. Macrofungi from *Abies cephalonica* forests and from other intermixed tree species (Oxya Mt., central Greece). Mycotaxon 2008; 104: 39-42.

[82]   Cailleux R, Diop A. Recherches préliminaires sur la fructification en culture du *Pleurotus eyngii* (Fr. ex DC.) Quélet. Rev Mycol 1976; 40: 365-388.

[83]   Cailleux R, Diop A, Joly P. Relations d'interfertilité entre quelques représentents des Pleurotes des Ombellifères. Bull Soc Mycol Fr 1981; 97: 97-124.

[84]   Joly P, Cailleux R, Cerceau M-T. La stérilité mâle pathologique, élément de la co-adaptation entre populations de champignons et de plantes-hôtes: modèle des Pleurotes des Ombellifères. Bull Soc Bot Fr-Actual 1990; 137: 71-85.

[85]   Bresadola J. Iconographia Mycologica. I. Mediolani: 1927.

[86]   Ferri F. Un medesimo substrato per piu specie fungine. Mushr Inf 1985; 3: 24-30.

[87]   Polemis E, Dimou DM, Tzanoudakis D, Zervakis GI. Annotated checklist of *Basidiomycota* (subclass *Agaricomycetidae*) from the islands of Naxos and Amorgos (Cyclades, Greece). Ann Bot Fenn 2012; 49: 145-161.

[88]   Polemis E, Dimou DM, Tzanoudakis D, Zervakis GI. Diversity of *Basidiomycota* (subclass *Agaricomycetidae*) in the island of Andros (Cyclades, Greece). Nova Hedwigia 2012; 95: 25-58.

[89]   Venturella G, Zervakis G, La Rocca S. *Pleurotus eryngii* var. *elaeoselini* var. nov. from Sicily. Mycotaxon 2000; 76: 419-427.

[90]   Venturella G. On the real identity of *Pleurotus nebrodensis* in Spain. Mycotaxon 2002; 84: 445-446.

[91]   Venturella G, Zervakis G, Saitta A. *Pleurotus eryngii* var. *thapsiae* var. nov. from Sicily. Mycotaxon 2002; 81: 69-74.

[92]   Ravash R, Shiran B, Alavi A-A, Bayat F, Rajaee S, Zervakis GI. Genetic variability and molecular phylogeny of *Pleurotus eryngii* species-complex isolates from Iran, and notes on the systematics of Asiatic populations. Mycol Progr 2010; 9: 181-194.

[93]   Rodriguez Estrada AE, del Mar Jimenez-Gasco M, Royse DJ. *Pleurotus eryngii* species complex: Sequence analysis based on partial EF1 a and RPB2 genes. Fungal Biol 2010; 114: 421-428.

[94]   Chinan VC, Venturella G. *Pleurotus eryngii* var. *elaeoselini*, first record from Romania. Mycotaxon (in press).

[95]   Pollack FG, Miller OK. Antromycopsis broussonetiae found to be the name of the imperfect state of *Pleurotus cystidiosus*. Mem New York Bot Gard 1976; 28: 174-178.

[96]   Moore RT. Deuteromycetes III. The other species of Antromycopsis. T Brit Mycol Soc 1984; 82: 377-380.

[97]   Guzman G, Bandala VM, Montoya L. A comparative study of teleomorphs and anamorphs of *Pleurotus cystidiosus* and *Pleurotus smithii*. Mycol Res 1991; 95: 1264-1269.

[98]   Pegler DN. *Pleurotus* (*Agaricales*) in India, Nepal and Pakistan. Kew Bull 1977; 31: 501-510.

[99]   Nair LN, Kaul VP. The anamorphs of *Pleurotus sajor-caju* (Fr.) Singer and *Pleurotus gemmellarii* (Inzeng.) Sacc. Sydowia 1980; 33: 221-224.

[100] Petersen RH, Krisai-Greilhuber I. An epitype specimen for *Pleurotus ostreatus*. Mycol Res 1996; 100(2): 229-235.

[101] Hilber O. Einige Aspekte aus der *Pleurotus ostreatus* Gruppe. Ceska Mykol 1977; 31: 142-154.

[102] Eger G, Li SF, Leal-Lara H. Contribution to the discussion on the species concept in the *Pleurotus ostreatus* complex. Mycologia 1979; 71: 577-588.

[103] Vilgalys R, Smith A, Sun BL, Miller OK. Intersterility groups in the *Pleurotus ostreatus* complex from the continental United States and adjacent Canada. Can J Bot 1993; 71: 113-128.

CHAPTER 3

The Genus *Pleurotus* in Italy and the Sicilian *taxa*

Giuseppe Venturella[*], Maria Letizia Gargano and Alessandro Saitta

Department of Agricultural and Forest Sciences, Università de Palermo, viale delle Scienze 11, I-90128, Palermo, Italy

Abstract: This chapter provides a brief description of species diversity and distribution of taxa belonging to the genus *Pleurotus* for each Italian region. For *taxa* that grow on the Sicilian territory and belong to the so-called "*Pleurotus eryngii* species-complex" ecological data are also reported.

Keywords: *Apiaceae, Basidiomycota, Cachrys ferulacea*, Distribution, Ecology, *Elaeoselinum asclepium* subsp. *asclepium, Eryngium campestre, Ferula communis* var. *communis*, Fungi, Intra-population variability, Italy, *Pleurotus, Pleurotus eryngii* var. *elaeoselini, Pleurotus eryngii* var. *eryngii, Pleurotus eryngii* var. *ferulae, Pleurotus eryngii* var. *thapsiae, Pleurotus nebrodensis*, Sicily, Systematic, *Thapsia garganica*.

INTRODUCTION

The first references on the systematics of the genus *Pleurotus* (Fr.) P. Kumm in Italy are made by imporant mycologists such as Saccardo and Baglietto [1,2]. The latter adopted the division of Fries which places the genus *Pleurotus* in the family *Tricholomataceae* recognizing three subgenera within the genus.

An updated list on the species diversity of the genus *Pleurotus* in Italy was provided under the coordination of University of Tuscia (Viterbo) by the Working Group for Mycology of the Italian Botanical Society [3]. Eleven *Pleurotus* taxa (including 3 varieties) were listed in the Checklist of Italian Fungi *i.e. Pleurotus calyptratus* (Lindbald) Sacc., *P. cornucopiae* (Paulet) Rolland, *P. dryinus* (Pers.) P. Kumm., *P. eryngii* (DC.) Quél. var. *eryngii*, *P. eryngii* var. *elaeoselini*

*Address correspondence to Giuseppe Venturella:** Department of Agricultural and Forest Sciences, Università of Palermo, vialedelle Scienze 11, I-90128, Palermo, Italy; Tel: +39 091 238 91 234; Cell: +39 329 615 60 64; E-mails: giuseppe.venturella@unipa.it; venturellagiuseppe1@gmail.com

Venturella, Zervakis & La Rocca, *P. eryngii* var. *ferulae* Lanzi, *P. eryngii* var. *thapsiae* Venturella, Zervakis & Saitta, *P. nebrodensis* (Inzenga) Quél., *P. opuntiae* (Durieu & Lev.) Sacc., *P. ostreatus* (Jacq.) P. Kumm. and, *P. pulmonarius* (Fr.) Quél. *P. calyptratus* is only reported from Trentino Alto Agide while *P. cornucopiae* shows a wider distribution being present in Abruzzo, Basilicata, Calabria, Campania, Emilia-Romagna, Lazio, Liguria, Lombardy, Marche, Piedmont, Apulia, Sicily, Tuscany, Umbria and Veneto.

A less extensive distribution characterizes the Italian population of *P. eryngii* var. *ferulae* which is reported for Basilicata, Calabria, Lazio, Lombardy, Apulia, Sardinia, Sicily and Umbria.

P. dryinus and *P. eryngii* var. *eryngii* are reported in many regions of Italy, while less common are *P. pulmonarius* and *P. opuntiae*, the latter reported only from Calabria and Sicily [3]. *P. ostreatus* is the most widely distributed species while *P. nebrodensis* and *P. eryngii* var. *thapsiae* are present only in Sicily.

Recently many reports of *P. nebrodensis* in Italy were re-identified as *P. eryngii* var. *elaeoselini*, a taxon with a much wider distribution in Europe than was previously assumed [4].

The infrastructure of the "*Pleurotus eryngii* species-complex" in Sicily is composed of the following host-specialized *taxa*: *P. eryngii* (DC.) Quél. var. *eryngii* growing on root residues of *Eryngium campestre* L., *P. eryngii* var. *ferulae* Lanzi, on *Ferula communis* L. var. *communis*, *P. nebrodensis* (Inzenga) Quél. on *Cachrys ferulacea* (L.) Calestani, *P. eryngii* var. *elaeoselini* Venturella, Zervakis & La Rocca on *Elaeoselinum asclepium* (L.) Bertol. subsp. *asclepium* and, *P. eryngii* var. *thapsiae* Venturella, Zervakis & Saitta on *Thapsia garganica* L.

P. eryngii var. *eryngii* is a common *taxon* growing in arid pastures, from early spring to autumn (Fig. **1a,b**).

P. eryngii var. *ferulae* is abundantly collected in garrigues, waste lands and arid pastures from September to the first ten days of November (Fig. **2a,b**).

Figure 1: a) *Eryngium campestre*; **b)** *P. eryngii* var. *eryngii*.

During winter the basidiomata appearance is scarce but, again, they become more abundant from the end of March until mid-June.

Both *P. eryngii* var. *eryngii* and *P. eryngii* var. *ferulae* are widespread throughout Sicily [5], mainly on calcareous soils. Particularly the former is collected from sea-level to 1300 m while the latter does not exceed altitudes of 1000 m.

Figure 2: a) *Ferula communis*; **b)** *P. eryngii* var. *ferulae*.

P. nebrodensis, growing in arid pastures, is a rare *taxon* and its fructification period is restricted from the middle of April to the first ten days of June.

The *P. nebrodensis* (Fig. **3a,b**) growing sites in Italy are located only on the Madonie area (northern Sicily) on calcareous soils from 1200 to 2000 m.

Collections of *P. nebrodensis* are also reported from Mount Etna, on soils of volcanic origin.

P. eryngii var. *ealeoselini* (Fig. **3c,d**), recently described as a new taxon [6], is widely distributed throughout Sicily from sea level to 1200 m in arid pastures on calcareous soils and it appears during the period March-May and October-November.

P. eryngii var. *thapsiae* (Fig. **3e,f**), recently described as a new taxon [7], is collected on calcareous soils, in arid pastures at elevations ranging from 0 to 1500 m, from the end of March until May.

In the case of *Pleurotus* systematics there has been controversy in the past, especially regarding taxa such as *P. nebrodensis*, *P. eryngii* var. *ealeoselini* and *P. eryngii* var. *thapsiae*.

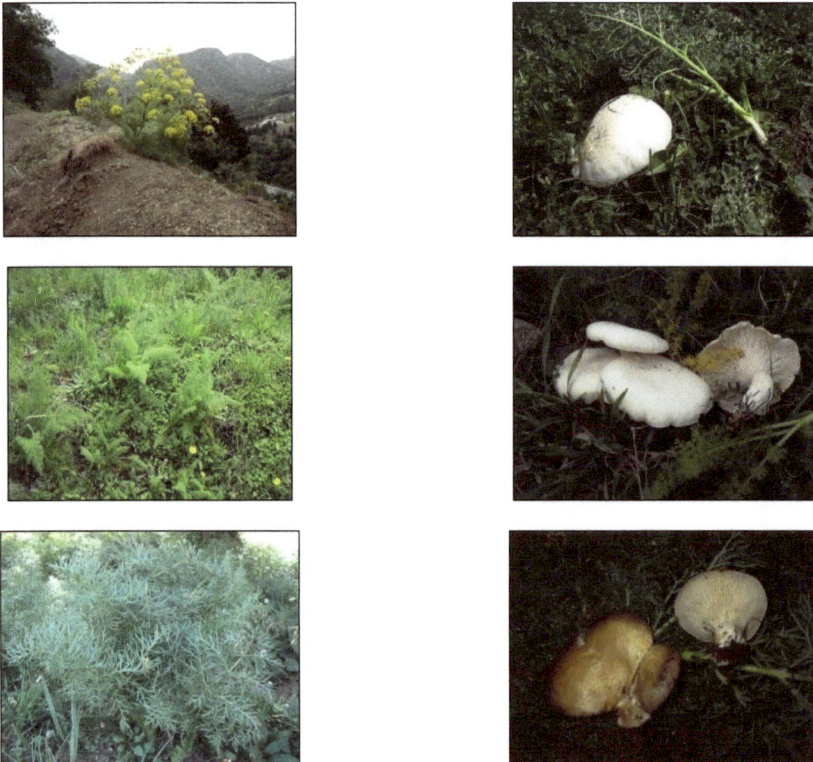

Figure 3: a) *Cachrys ferulacea*; **b)** *P. nebrodensis*; **c)** *Elaeoselinum asclepium* subsp. *asclepium*; **d)** *P. eryngii* var. *ealeoselini*; **e)** *Thapsia garganica*; **f)** *P. eryngii* var. *thapsiae*.

These taxonomic ambiguities are due to initial misidentifications, absence of species typification, and significant influence exercised by the environment and substrate on the morphological characters of each *taxon*.

The separation among the *Pleurotus* taxa based on macro- and micro-morphological characters was also confirmed by isozyme and molecular techniques [8].

Both isozyme and RAPD-PCR analysis showed that *Pleurotus* isolates growing on *Cachrys ferulacea* formed a cluster with relatively high statistical support, presenting increased genetic distances with other populations. Intra-population variability within this group was low, which is indicative of the large coherence among the fungal individuals associated with *Cachrys ferulacea*. All other strains were positioned within the larger *P. eryngii* group, which was further divided into four main clusters corresponding to *Eryngium*, *Ferula*, *Elaeoselinum* and *Thapsia* hosts.

ACKNOWLEDGEMENT

Declared none.

CONFLICT OF INTEREST

The author(s) confirm that this chapter content has no conflict of interest.

REFERENCES

[1]   Saccardo PA. Flora italica cryptogama. Fungi Hymeniales. Pars I. Rocca San Casciano 1386; pp. 1915-6.
[2]   Baglietto C. Saggio di flora micologica analitica con particolare riguardo per la flora ligustica. Note sulla biologia e sulla sistematica, discussione di specie rare o critiche, Funghi superiori. Italy: Genova 1972; pp. 526.
[3]   Onofri S. Check-list of Italian fungi. Basidiomycota. Delfino Editore, Italy: Sassari 2005.
[4]   Chinan VC, Venturella G. *Pleurotus eryngii* var. *elaeoselini*, first record from Romania. Mycotaxon (in press).
[5]   Venturella G. A check-list of Sicilian fungi. Bocconea 1991; 2: 1-221.
[6]   Venturella G, Zervakis G, La Rocca S. *Pleurotus eryngii* var. *elaeoselini* var. nov. from Sicily. Mycotaxon 2000; 76: 419-27.
[7]   Venturella G, Zervakis G, Saitta A. *Pleurotus eryngii* var. *thapsiae* var. nov. from Sicily. Mycotaxon 2002; 81: 69-74.

[8]   Zervakis G, Venturella G, Papadopoulou K. Genetic polymorphism and taxonomic infrastructure of the *Pleurotus eryngii* species-complex as determined by RAPD analysis, isozyme profiles and eco-morphological characters. Microbiology 2001; 147: 3183- 94.

Send Orders for Reprints to reprints@benthamscience.net

CHAPTER 4

An Outline of the Madonie Mountains (Northern Sicily): A Center of Diversity for *Pleurotus* Species

Giuseppe Venturella[*] and Alessandro Saitta

Department of Agricultural and Forest Sciences, Università de Palermo, viale delle Scienze 11, I-90128, Palermo, Italy

Abstract: The territory of the Madonie Mountains is considered as a biodiversity hotspot in the Mediterranean Basin. Since long distinguished botanists who have studied the vegetation of this particular territory were fascinated by the high level of species diversity. In this chapter, an overview of the geographical, geological, soil, climatic and vegetation features of the area is provided. Madonie territory is an important center of diversity for fungi and in particular for species belonging to the genus *Pleurotus* growing on roots of plants of the family *Apiaceae*.

Keywords: *Apiaceae*, *Basidiomycota*, Biodiversity, *Cachryetum ferulaceae* subass. *cachryetosum*, *Cachrys ferulacea*, Checklist, *Elaeoselinum asclepium* subsp. *asclepium*, *Fagus sylvatica*, *Ferula communis* var. *communis*, Fungal diversity, Madonie Mts, Mediterranean Basin, *Pleurotus*, *Pleurotus eryngii* var. *elaeoselini*, *Pleurotus eryngii* var. *eryngii*, *Pleurotus eryngii* var. *ferulae*, *Pleurotus eryngii* var. *thapsiae*, *Pleurotus nebrodensis*, Sicily, Vegetation landscape.

INTRODUCTION

The territory of Madonie Mts covers an area of 40000 hectares and is located in the northern part of Sicily, about 5 km from the Tyrrhenian coast.

The territory extends between the Pollina and the Northern Imera rivers; it is characterized by high mountains such as Pizzo Carbonara (1979 m), Pizzo Antenna Grande (1977 m), Pizzo Palermo (1955 m), Monte San Salvatore (1912 m), Monte Ferro (1906 m), Monte Quacella (1869 m), Monte Mufara (1865 m), Monte dei Cervi (1794 m), and Pizzo Dipilo (1385 m).

*Address correspondence to Giuseppe Venturella: Department of Agricultural and Forest Sciences, Università of Palermo, vialedelle Scienze 11, I-90128, Palermo, Italy; Tel: +39 091 238 91 234; Cell: +39 329 615 60 64; E-mails: giuseppe.venturella@unipa.it; venturellagiuseppe1@gmail.com

Maria Letizia Gargano, Georgios I. Zervakis and Giuseppe Venturella (Eds)

The name "Madonie" refers to the mountainous system to which the vernacular name "Nebrodi" also applies.

Nowadays the Nebrodi Mountains are geographically included in the province of Messina (eastern Sicily).

Many scientific binomials used for the description of plants, animals and fungi living on the Madonie area employ the epithet *nebrodensis*. The improper use of such specific epithet was introduced by the German writer T. Fischer [1] in his work on the geography of Sicily that was published earlier than the monography "Flora der Nebroden" [2] which deals exclusively with the Madonie area.

M. Lojacono Pojero, in the introduction of "Flora Sicula" [3], analyzed the geographical mistake introduced by Fischer. He reported that the name "Madonie" refers to the true Nebrodi and that it is derived from the portion of the mountain range that corresponds to the old fief "Madonia" [4]. The territory of Madonie shows a merely mountainous morphology with a calcareous dolomitic landscape characterized by peaks, cliffs, rocky ridges, craggy walloons and quarzarenite and flysch areas.

Strongly eroded areas represented by marly clays and scaly clays are also included throughout the Madonie area. The Madonie territory shows a very heterogeneous composition with outcrops that date back to the Upper Triassic and Quaternary subdivided in well-defined and laid upon stratigraphic-structural units termed "complex" [5,6]. Referring to the USDA Soil Taxonomy [7] and to the studies of Sicilian pedologists [8] the main soil cartographic units in the Madonie area are: a) Lithic xerorthents, Rock outcrop and Lithic haploxerolls; b) Lithic xerorthenths, Rock outcrop and, Typic and/or Lithic xerochrepts; c) Typic xerorthents, Typic and/or Vertic xerochrepts, and, Typic and/or Vertic xerofluvents and/or Typic chromoxererts and/or Typic pelloxererts; d) Typic xerorthents, Typic xerochrepts, Typic haploxeralfs, Typic xerochrepts, Typic haploxeralfs, and Typic and/or Lithic xerorthents.

The climate is characterized by cold winters with snowfalls and a long xerothermic summer periods. The average rainfall exceeds 800 mm and the rainy days are mainly concentrated in autumn and winter but sometimes they also appear during spring and summer. The mean annual temperature is lower than 18°C.

The north facing slopes are exposed to humid air currents that improve the water supply causing peculiar microclimatic conditions of oceanic type.

The Madonie territory shows a very high degree of diversity displayed by about 1600 *taxa* of the vascular flora and a high endemism-rate.

According to Pignatti [9] four vegetation belts could be recognized on the Madonie Mts: Mediterranean-arid, Mediterranean, Colchic and Subatlantic. Above the Subatlantic belt the vegetation is characterized by spiny shrubs of *Astragalus nebrodensis* (Guss.) Strobl.

The calcareous and triassic dolomite stony slopes are characterized by shrubby vegetation with many herbaceous species particularly in the higher mountains such as Pizzo Carbonara and Monte dei Cervi [10].

In the glades of beech woods in these mountainous areas, pastures of *Cachrys ferulacea* (L.) Calestani are present from 1400 to 1900 m [11] (Fig. **1**).

Figure 1: Pasture of *C. ferulacea.*

Cachrys ferulacea (L.) Calest. (*Apiaceae*) is a N.-E. Mediterranean-Turanic, perennial, hemicryptophyte scapose plant, 3-15 dm. The stem is erected, striate, branched. The leaves are 2-5 dm, 3-5 pennatosette, completely divided into linear segments (1 × 10-30 mm). Rays of the umbel 8-15, bracts and bracteoles linear. The flowering period is between June and August. The petals, yellow in color, are beaming with bent apex. The fruit, ranging in size from 15 to 30 mm, has ribs prolonged into wings.

C. ferulacea is distributed in central-southern Italy, in dry meadows on limestone from 0 to 2000 m a.s.l. The plant reaches in Sicily the westernmost border of its European distribution. Locally named "basilisco or basiliscu", is distributed in the main mountain chains of Sicily (Peloritani, Nebrodi, Madonie, Monte Cammarata, Rocca Busambra, Pizzuta and Etna). *C. ferulacea* is used as forage due to the intense and pleasant aroma conveyed to milk and cheese produced locally.

C. ferulacea features xerophilous grassland characterized by *Artemisia alba* Turra, *Asphodeline lutea* (L.) Rchb., *Astragalus depressus* L., *Bonannia graeca* (L.) Halácsy, *Calamintha nepeta* (L.) Savi, *Centaurea parlatoris* Heldr., *Erysimum bonannianum* C. Presl, *Euphorbia rigida* M. Bieb., *Festuca circummediterranea* Patzke, *Helianthemum croceum* (Desf.) Pers., *Hieracium macranthum* (Ten.) Zahn, *Knautia calycina* (C. Presl) Guss, *Medicago lupulina* L. subsp. *cupaniana* (Guss.) Nyman, *Melica ciliata* L. subsp. *ciliata*, *Micromeria juliana* (L.) Benth. ex Rchb., *Rosa sicula* Tratt., *Scutellaria rubicunda* Hornem. subsp. *linneana* (Caruel) Rech. fil, *Sesleria nitida* Ten. subsp. *sicula* Brullo & Giusso, *Stachys germanica* L. var. *dasyanthes* (Raf.) Arcang., *Thlaspi rivale* C. Presl, *Vicia glauca* C. Presl, *Viola nebrodensis* C. Presl, *etc.*

The slopes of the Madonie Mts. are dominated by pastures of *C. ferulacea* which are considered to form the last stage of degradation of beech (*Fagus sylvatica* L.) forests and soils destroyed mainly because of over grazing and human activities.

At the same altitudes, but in the north facing slopes, *Fagus sylvatica* L. woods are localized on calcareous and siliceous substrata (Fig. **2**).

Figure 2: Partial view of *F. sylvatica* forest on the Madonie Mts.

The most extensive beech woods are present on Monte Mufara, Pizzo Antenna Grande and in the neighbouring regions of Monte dei Cervi, Monte Daino and Monte San Salvatore.

Fagus sylvatica woods are restricted in a vegetational belt ranging from 1100 to 2200 m. The survival of beech forests in the dry climatic conditions of Sicily is mainly attributed to the presence of snow during the period January-March and, during summer, to occasional incidences of fog and dew [12-14]. The beech woods are located on calcareous and siliceous substrata, on soils with very limited thickness, and submitted to very dry climatic conditions.

The classification of beech forests is mainly referred to the *Anthrisco-Fagetum* described by Hofmann [15]. This author distinguishes two sub-associations: *Anthrisco-Fagetum luzuletosum* on siliceous-arenaceous substrates and *Anthrisco-Fagetum aceretosum* on calcareous-dolomite substrates.

In the Madonie area two different vegetation types could be distinguished: a) beech woods located on northern slopes and belonging to *Geranio versicoloris-Fagion* alliance (altitudinal belt ranging from 1200 to 1975 m), and b) scattered nuclei of coppiced beech woods.

The fragmentation of beech woods in the Madonie territory is mainly caused by the extreme environmental conditions of Sicily in which plants are adapted to grow at the southernmost border of their area of distribution. Important stress factors are also the irrational silvicultural treatements and grazing and the increasing human pressure on habitats too (Fig. **3**).

In the glades of beech woods and at the border of such interesting forestry areas, the plant association named *Cachryetum ferulaceae* Raimondo 1980, subass. *cachryetosum* Brullo 1984 belonging to *Cerastio-Astragalion nebrodensis* Brullo 1984 could be observed. This is a mesophilous and nitrophilous association linked to rocky, karstic habitats, with very primitive soils [16].

Figure 3: Horses grazing in a clearing adjacent to the beech forest.

In the Pizzo Cerasa and Vallone San Nicola area, on acid soils distributed from 1400 to 1600 m, a peculiar type of woods characterized by *Quercus petraea* (Mattuschka) Liebl. is present.

It is characterized by secular trees of oak mixed with beech trees, *Acer pseudoplatanus* L. and a dense and intricate shrubby underbrush of *Ilex aquifolium* L.

Isolated plants of *Q. cerris* L. together with broad-leaved oaks belonging to the *Q. pubescens* Willd s. l. are present too.

Mixed woods of *Quercus ilex* L., *Q. suber* L. and *Q. pubescens.* s. l. with few plants of *Acer campestre* L. and *Fraxinus ornus* L. are also present on calcareous and siliceous substrata below 1400 m (Fig. **4**).

Figure 4: Mixed oak woods on the top of a hill.

On siliceous substrata, below 700 m, *Quercus suber* is dominant and sometimes forms mixed woods with *Q. dalechampii* Ten., *Q. congesta* Presl. and, sporadically, with *Q. ilex* (Fig. **5**).

Quercus ilex is a fundamental element of the Mediterranean forest particularly because of its adaptability to different pedoclimatic conditions.

On the Madonie Mts, *Q. ilex* shows its optimum growth in the altimetrical belt located from 400 to 1200 m but it could also reach the subatlantic belt and come into contact with *F. sylvatica* [11] (Fig. **6**).

At the same altitude, Castanea sativa Miller woods mixed with Q. ilex, Q. pubescens, Q. suber, Fraxinus ornus and F. oxycarpa Bieb. are present (Fig. **7**).

In the entire Madonie area woods are utilized to obtain firewood, timber and charcoal. These activities together with grazing and fire are the main reasons why woods gradually declined both in their structures and floristic compositions.

Figure 5: A degraded cork oak wood with an undergrowth of *Cistus* spp.

Consequently the woody areas underwent a strong reduction so that most of the Madonie territory is at present characterized by more or less stripped areas where a herbaceous and shrubby vegetation is present, which is mainly used for grazing.

Figure 6: A *Q. ilex* forest degrading on a slope.

Figure 7: A coppiced chestnut wood.

Along the rivers, maquis and shrubs are characterized by *Populus alba* L., *Salix alba* L., *Fraxinus oxycarpa* and *Sambucus nigra* L. in the mountain part of the Madonie Park, while *Salix pedicellata* Desf., *Tamarix africana* Poiret, *Populus nigra* L. and, again, *S. alba* are present at lower altitudes.

Finally the thermo-mediterranean belt is characterized by *Olea europaea* L. var. *sylvestris* Hoffmgg. et Link, *Myrtus communis* L. and *Pistacia lentiscus* L., and maquis with *Cistus creticus* L., *C. salvifolius* L. and *Erica arborea* L.

Cultivated fields mainly include sown lands cultivated with cereals and leguminous, olive groves, almond- and ash-trees orchards.

Pinus halepensis Miller, P. pinea L., P. nigra Arnold, Cedrus atlantica (Endl.) Carrière, Cupressus sempervirens L., C. arizonica Green, C. macrocarpa Hartweg, Abies alba Miller, A. cephalonica Link, Pseudotsuga menziesii Franco var. menziesii, Robinia pseudoacacia L. and Eucalyptus camaldulensis Dehnh. are the main introduced exotic species in the reafforested areas (Figs. **8**, **9**).

As regards fungi, 614 *taxa* (63 classified within *Ascomycota* and 551 within *Basidiomycota*) including 25 varieties and 8 forms, belonging to 200 genera included in 79 families were recorded [17].

Figure 8: Top view of a Cedar tree planting.

Figure 9: Reforestation with *Pinus* sp.

The richest areas for fungi are concentrated in the central part of the Madonie territory and mainly correspond to the more important woody areas.

All *Pleurotus* species listed in the checklist of Sicilian fungi [18] are also reported from the territory of the Madonie. All five taxa of the genus *Pleurotus*, so far described for Sicily [*P. eryngii* (DC.) Quél. var. *eryngii*, *P. eryngii* var. *ferulae* Lanzi, *P. nebrodensis* (Inzenga) Quél., *P. eryngii* var. *elaeoselini* Venturella, Zervakis & La Rocca and, *P. eryngii* var. *thapsiae* Venturella, Zervakis & Saitta], growing on rotten roots of the *Apiaceae*, in different periods of the year can be found in meadows and pastures of the Madonie mountains.

In addition to their ecological value, *Pleurotus* mushrooms growing on the roots of *Apiaceae* are known to the local population for their excellent organoleptic properties and are harvested, marketed and used by restaurants throughout the year.

ACKNOWLEDGEMENT

Declared none.

CONFLICT OF INTEREST

The author(s) confirm that this chapter content has no conflict of interest.

REFERENCES

[1]   Fischer T. Beiträge zur physichen Geographie der Mittelmeerländer, Leipzig 1877.
[2]   Strobl PG. Die Dyalipetalen der Nebroden, Siziliens. Verh K K Zool-Bot Ges Wien 1903; 53: 434-558.
[3]   Lojacono Pojero M. Flora Sicula. I-II-III. Tipografia Virzì: Palermo 1888-1908.
[4]   Raimondo FM. On the natural history of the Madonie mountains. Webbia 1984; 38: 29-61.
[5]   Abate B, Catalano R, D'Argento B, *et al*. Facies sedimentarie e rapporti strutturali nelle Madonie orientali (con carta geologica). Palermo: Guida alla Geologia della Sicilia occidentale 1982; pp. 49-52.
[6]   Lentini F, Vezzani L. Carta geologica delle Madonie (Sicilia centro-settentrionale), Italy: Firenze 1978.
[7]   Soil Survey Staff. Soil Taxonomy USDA, Soil Conservation Service. Agric. Handbook, 436. Washington DC. 1975
[8]   Fierotti G, Dazzi C, Raimondi S. Commento alla carta dei suoli della Sicilia. In: Fierotti G, Carta dei Suoli della Sicilia, Assessorato Territorio ed Ambiente, Palermo: Italy 1988.
[9]   Pignatti S. I piani di vegetazione in Italia. Giorn Bot Ital 1979; 113: 411-28.
[10]   Raimondo FM, Gianguzzi L, Schicchi R. Carta della vegetazione del massiccio carbonatico delle Madonie (Sicilia centro-settentrionale). Quad Bot Ambientale Appl 1994; 3(1992): 23-40.
[11]   Raimondo FM. Carta della vegetazione di Piano Battaglia e del territorio circostante (Madonie, Sicilia). CNR Programma finalizzato "Promozione della Qualità dell'Ambiente". s. AQ/1/89. Roma 1980; pp. 43.
[12]   Brullo S. Contributo alla conoscenza della vegetazione delle Madonie. Boll Accad Gioenia Sci Nat Catania 1993; 16(322): 351-420.
[13]   Brullo S, Guarino R, Minissale P, Siracusa G, Spampinato G. Syntaxonomical analysis of the beech-forests from Sicily. Ann Bot 1999; 57: 121-32.
[14]   Ferro G, Coniglione P, Oliveri S, Scuderi M, Grasso S. Osservazioni fitosociologiche sugli aggruppamenti boschivi di Sicilia. Atti Accad Gioenia Sci Nat Catania 1980; s. 4, 13(9): 137-41.

[15]   Hofman A. Il faggio in Sicilia. Flora e Vegetatio Italica, 2. Gianasso, Italy: Sondrio1960.

[16]   Brullo S, Cormaci A, Giusso del Galdo G, *et al*. A syntaxonomical survey of the Sicilian dwarf shrub vegetation belonging to the class Rumici-Astrgaletea siculi. Ann Bot 2005; 5: 57-104.

[17]   Venturella G, Saitta A, La Rocca S. A check-list of the mycological flora of Madonie Park (North Sicily). Mycotaxon, 2000 Ltd; pp. 246.

[18]   Venturella G. A check-list of Sicilian fungi. Bocconea 1991; 2: 1-221.

CHAPTER 5

Historical Remarks, Original Description and Last Advances on *Pleurotus nebrodensis*

Maria Letizia Gargano and Giuseppe Venturella*

Department of Agricultural and Forest Sciences, Università de Palermo, viale delle Scienze 11, I-90128, Palermo, Italy

Abstract: Giuseppe Inzenga, the scientist who first described *Pleurotus nebrodensis*, was one of the major mycologists of the second half of the 19th century. This chapter describes the main biographical aspects of the life of Inzenga and has contribution to science with particular reference to mycology. The chapter also includes notes on the social and academic context of the mid-1800s, which led Inzenga to the original description of *P. nebrodensis*, the analysis of the critical aspects of nomenclature and the logical path that led us to the recent designation of a *P. nebrodensis* epitype. An updated description of the taxon and the discovery of the *loci classici* of *P. nebrodensis* collections is also reported.

Keywords: *Agaricus nebrodensis*, *Apiaceae*, *Basidiomycota*, *Cachrys ferulacea*, Centurie, Description, Elias Fries, Fungi, Giuseppe Inzenga, Iconography, *Loci classici*, Madonie Mts., Madonie Regional Park, Pastures, *Pleurotus*, *Pleurotus eryngii* var. *elaeoselini*, *Pleurotus nebrodensis*, Sicily, Taxonomy, Vegetation.

INTRODUCTION

The biographies of Giuseppe Inzenga (1816?-1887), son of Gaetana Angles and Pompeo Inzenga, the scientist (Fig. **1**) who first described *Pleurotus nebrodensis* [sub: *Agaricus nebrodensis* Nob.] are rather brief and in some cases giving contradictory information, such as his birth date. According to Savastano (1915) [1], Inzenga was born in Palermo in 1816, whereas in Muccioli's opinion (1998) [2], a year earlier.

Inzenga, after having studied the classics at Saint Anna of Jesuit School, and during his studies at the Real University of Palermo, showed an outstanding

---

*****Address correspondence to Giuseppe Venturella:** Department of Agricultural and Forest Sciences, Università of Palermo, vialedelle Scienze 11, I-90128, Palermo, Italy; Tel: +39 091 238 91 234; Cell: +39 329 615 60 64; E-mails: giuseppe.venturella@unipa.it; venturellagiuseppe1@gmail.com

eclecticism by dedicating himself to studying medicine and afterwards (following the advice of Ruggiero Settimo), agriculture, physics and mathematics.

After obtaining a degree in each discipline, he earned a "diploma" in land-surveying allowing him to become Associate Professor of Agriculture on December 5th, 1842.

Figure 1: Portrait of Giuseppe Inzenga with signature.

On April 19th, 1844, Ruggiero Settimo bestowed him the position of Permanent Director of the Agricultural Institute of Castelnuovo (Fig. **2**), founded by Carlo Cottone, Prince of Castelnuovo, located in his "country house" in the Colli Plains, adjacent to the Verdura Theater of Palermo. Carlo Cottone wrote his first testament for the foundation of the Agricultural Institute at the Colli on July 20th, 1827, although his work had begun some years before.

After the death of Cottone, the construction of the Institute continued on account of Ruggiero Settimo, executor. Before assuming the role of director, Inzenga traveled extensively throughout Sicily in order to become acquainted with the agricultural environment [3].

On October 20th, 1860, Inzenga was nominated Professor of Agriculture at the University of Palermo, and in 1867, Chair of Economy and Rural Assessment.

A file about Giuseppe Inzenga, as a member of the Georgofili Academy, the Gioenia Academy of Catania, the Acclimation Society and many other agricultural Institutes, is conserved at the Public Central Archive in Rome.

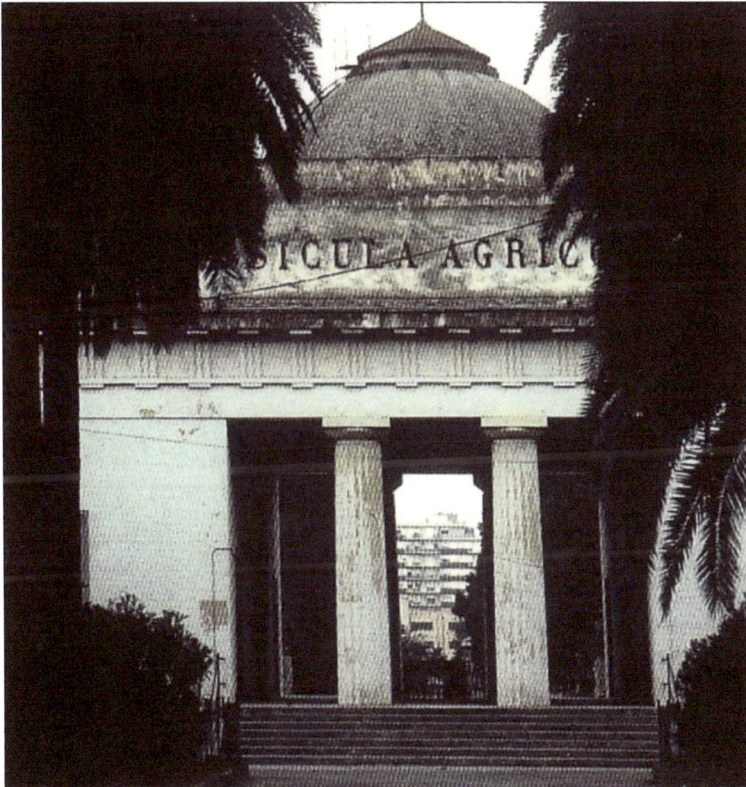

Figure 2: The Gymnasium of the Istituto Agrario Castelnuovo, which houses the library, where the unpublished material of Inzenga was found.

In that file, a proposal to nominate the Palermitan scholar as a Scholastic Councilman for the Province of Palermo following a transfer to the University of

Rome by Professor Stanislao Cannizzaro is found (Public Central Archive in Rome, Ministry of Education, Personal 1860-1880, "Inzenga, Giuseppe" file).

Other brief biographical annotations are contained in publications of many authors [4-13]. The varied interests of Inzenga are proven in numerous scientific publications of different research fields, including Agronomy, Agrarian Industries, Botany, Mechanical Agronomy, Arboriculture and Plant Pathology. Among the various contributions, those relative to the cultivation methods of sugar cane, carob and saffron are mentioned. Inzenga also wrote a practical handbook on the cultivation of the vine plant and published verses in Sicilian dialect under the pseudonym Eugenio Savarese.

Another research field of significant interest, strangely little or not at all cited in the biographies about Inzenga, is that of mycology and macromycetes in particular. From 1865, these subject areas became a fundamental part of his scientific research.

The results of Inzenga's numerous studies were published in the journal "Nuovi Annali di Agricoltura Siciliana" (New Annals of Sicilian Agriculture) of which he was not only the founder, but also the director until his death on October 30th, 1887.

During the scientific meeting of the Society of Natural and Economic Sciences of Palermo held on December 27th, 1874, Inzenga presented a lecture entitled "Considerations on edible and poisonous fungi in Sicily" in which beyond supplying information about the progress of his mycological studies, he also demonstrated that *50 species of fungi, ten of which were entirely new to science, furnishing the respective diagnoses and colored drawings made by him using the actual mushrooms as models.*

The renewed interest from the scientific community towards mycological studies around the second half of the 1800's aided Inzenga's investigations. In fact, one can add Persoon's and Elias Fries' botanical volumes to the important descriptive works by Schaeffer, Paulet and Bulliard written toward the end of the 1700s. Modern taxonomy of fungi began after Elias Fries published his work *Systema mycologicum.* After these publications, the iconographic works of Briganti, Viviani, Venturi and Vittadini [14-17] followed. Inzenga acquired the majority of the afore mentioned books for the Castelnuovo Institute's Library.

After beginning to observe the diversity of fungi of *Palermo's surrounding country*, and having spoken at the May 13th, 1877 meeting during a presentation entitled "Varied Production of Fungi in Sicily", Inzenga realized the affinity of the Sicilian fungi with their European counterparts, and reported the discovery of new and rare fungal species. He also stated, that *which is remarkable is the similarity between the species in Northern Europe with those of Southern Europe, as well as in neighboring Africa, as noted in old and modern mycological works, which many have studied until now.*

Inzenga did not limit himself only to the study of fungi from a scientific point of view, but he also published elaborated studies on mushrooms' alimentary qualities and risks caused by the careless consumption of toxic and poisonous species.

For example, in the 138th issue of the Annals of Sicilian Agriculture published in 1882, Inzenga included an article entitled "Be careful when eating mushrooms" which contained useful advices still valid today, for the correct identification of mushroom fungi. Inzenga wrote: *Taking into consideration that our markets have already started selling these wild and delicious products, which the ancient food experts called "Food for the Gods", for what it is worth, we would like to warn our consumers to be careful as regards the purchase of muhrooms sold in bulk, as it was done in the past, without being able to distinguish the truly harmless edible species from other suspicious and poisonous, which at times because of lack of know-how from mushroom collectors, they could be found in our markets. Sometimes in certain periods, poisonous species appear and grow alongside with known edible species and because of their similar external appearance, the former are confused with the latter. In any case, one needs to consider that when regarding mushrooms, also for those being well known and safe to eat, like our Agaricus nebrodensis, fungi from Cachrys ferulacea (Apiaceae) or from the Madonie, which has been present in the markets of Palermo for a few years during the latter part of spring, the requirement to be fresh and in an early stage of development should not be ignored because the most innocent and edible mushrooms once having matured and become overripe, if they are not poisonous enough to kill, they are still difficult to digest and are capable of producing serious gastric upsets, which would not be the most pleasant situation in the world.*

Noteworthy are also the papers on the use of atropine as a "Remedy for mushroom poisoning" and "Antidote for poisonous mushrooms". The latter refers to the method of *an energetic vomiting* which as he says *it is sufficient to empty the stomach of all mushrooms eaten for preventing further unpleasant consequences.*

Along with Professor Emanuele Paternò, the director of the Chemistry Laboratory at the University of Palermo, Inzenga was appointed by the City Hall as the person to conduct a survey on a group of dried mushrooms responsible for several cases of intoxication among the population of Palermo.

On this occasion as well, Inzenga did not miss the opportunity to inform the citizens of Palermo about the risks deriving from an improper use of fungi by publishing an article in the Annals in 1886 entitled "Dried Fungi".

Articles on the "Artificial Cultivation of mushrooms", such as Agaricus campestris, successfully performed in the garden of the Fossa della Garofola of His Royal Highness the Duke d'Aumale by the agile flower gardener Giosuè Terzi can also found in the Annals.

During the period of resplendence for mycology in Sicily, Inzenga found good collaborators in Giuseppe Scalia, who investigated a large number of fungi from the province of Catania, and in Francesco Minà Palumbo, a naturalist and medical doctor from Castelbuono (a town of Madonie, N.Sicily), whose methodical research on fungi from the Madonie constituted a useful reference point for Inzenga during the time of his writing of the two Centurie about Sicilian fungi [18].

Nevertheless, the references in the Centurie to the Minà Palumbo collections are limited to only nine fungi which Inzenga affirmed to have described on the basis of a sample or of a drawing furnished by the Madonie scholar. Owing to a correspondence with Professor Giuseppe De Notaris, from the University of Genova, who provided *teachings, advices and dried specimens*, Inzenga initiated his project of the Centurie of Sicilian fungi.

The fungi included in the Centurie were either personally collected by Inzenga in the surrounding areas of Palermo or received by correspondents and friends, including the Baron Nicolò Turrisi Colonna and the aforementioned Francesco Minà Palumbo.

From a total of 200 *taxa*, 101 (99 species and 2 varieties), were published in the first Centuria, and 99 in the second. For each specimen observed, he included taxonomic annotations, some morphological characteristics and, if noted, the names in Sicilian dialect. In particular, there were 57 references to traditional uses, principally regarding foodstuffs, and 45 citations in the dialect used throughout the Sicilian countryside. The Centurie were annotated with 18 colored tables of which eight were included in the first volume along with 15 drawings, while the other 10 tables appeared in the second volume and corresponded to 21 drawings of fungi. In Inzenga's mycological works, references to authoritative characters of the time, to which the Palermitan scholar dedicated several species of fungi, were not lacking.

Among those included were *Agaricus bertoloni* dedicated to Antonio Bertoloni, Professor Emeritus of Botany at the University of Bologna and *Agaricus gemmellari* taken from Gaetano Giorgio Gemmellaro, Professor of Geology at the University of Palermo.

In addition, the species *Agaricus gussonii* was dedicated to his *friend, teacher and patron*, Giovanni Gussone, and another one to the musician Vincenzo Bellini *(Boletus bellini) the day that his ashes return to Catania.*

There is, without doubt, a dedication to *the master of living mycology*, Elias Fries *(Boletus friesii)*, to the Roman mycologist Matteo Lanzi *(Boletus lanzi)* and to *his friend and colleague*, Agostino Todaro *(Polyporus todari)*, director of the Botanical Gardens of Palermo.

Elias Fries was a fundamental reference point for Inzenga who, initially through De Notaris, and then directly from Palermo, sent numerous *exsiccata* and descriptions to the distinguished Swedish mycologist. Several herbarium samples and their respective drawings are found at the "Herb. Fries" of Botanical Museum of Uppsala (Sweden), while nine letters, (sent between 1869 and 1877 with "Joseph Inzenga's signature") are kept at Universitetbibliotek Handskriftsavdelningen in the same Swedish town. In another letter, dated March 13th, 1878 and addressed to Theodor Magnus Fries, son of Elias, Inzenga expressed *his utmost sorrow regarding the news of the death of the Swedish scientist.*

The project of Sicilian fungi collection in the Centurie started in 1866 with the publication entitled *"New Species of fungi and others discovered for the first time in Sicily by Professor Giuseppe Inzenga"*, which appeared in the Natural and Economical Sciences journal (Fig. **3**).

GIORNALE

DI

SCIENZE NATURALI ED ECONOMICHE

PUBBLICATO

PER CURA DEL CONSIGLIO DI PERFEZIONAMENTO

ANNESSO

AL R. ISTITUTO TECNICO DI PALERMO

VOLUME I.

PALERMO
STABILIMENTO TIPOGRAFICO DI FRANCESCO LAO
Via del Celso n. 31.

1866.

Figure 3: The inside cover of the first volume of the journal in which Inzenga published the first three species of his Centuries.

In Volume I of the same journal, dated April 1st, 1866, the first three species, *Hydnum notarisii, Agaricus gussonii* and *Agaricus bertoloni* were dedicated to Professor Giuseppe De Notaris (University of Genova), to the Italian botanists Giovanni Gussone and Antonio Bertoloni respectively, and were published following a brief introduction on the purpose of his study.

The publication of the first one hundred species finished in 1868 with a description of the last two species, conclusions, an alphabetical index, an index with Italian names and names in dialect used in various provinces, an index of Sicilian names in dialect and pertinent tables of some of the fungal species cited in the text.

The publication of the other one hundred species started in 1869 and was completed two years later. The two Centurie were then published in a two volume edition by Typography Lao of Palermo (Figs. **4 a,b**).

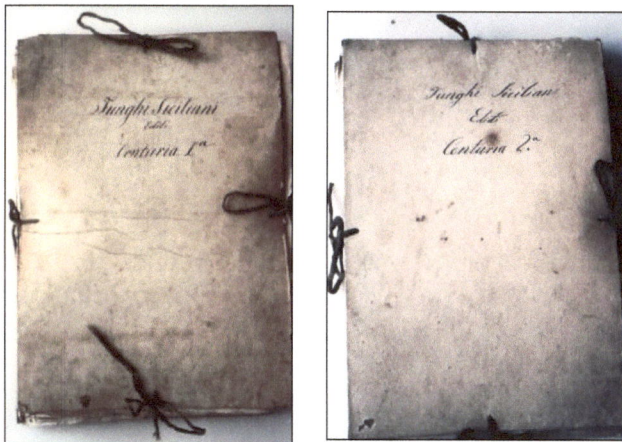

Figure 4: The hardcovers with strings of the herbarium packages containing the unpublished drawings and descriptions of Inzenga's mushrooms. **a)** I Centuria; **b)** II Centuria.

A significant amount of unidentified material was to constitute the basis for the publication of a third Centuria, which was never realized due to excessive printing costs. Inzenga sent a letter on December 23th, 1885 to mycologist Jean Baptiste Barla from Nice (France) stating *the third Centuria of my Sicilian fungi happens to be beautiful and complete not only for the drawings, but also for the text. However, since there are many more tables than those included in the two preceding Centurie, and considering that the cost of lithography is outrageously high, I had to postpone its publication. I hope this delay is temporary and that I will be able to show it to you soon.*

Inzenga [19] exposed the contents of a manuscript of 30 species of fungi from Sicily he had recently studied and described, with a corresponding table which contained two figures of *Agaricus nebrodensis* and one of Agaricus ferulae (originally described by Dr. Matteo Lanzi in Rome), a mushroom also found by Inzenga in Sicily (Fig. **5**). The two species were very similar in appearance, but they can be easily distinguished by expert mycologists on the basis of some key features.

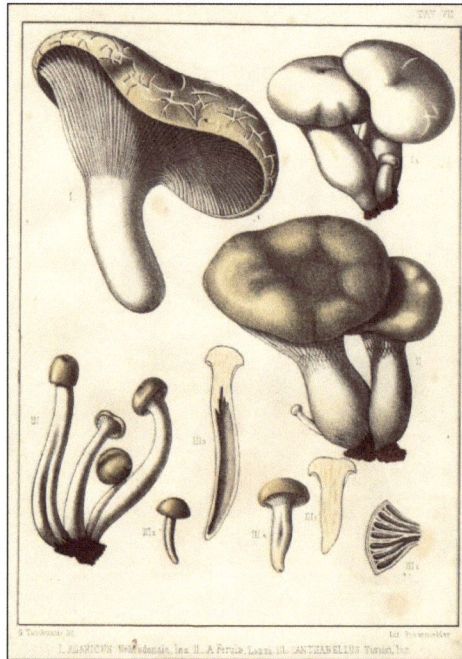

Figure 5: A table included in the I Centuria depicting *Pleurotus eryngii* var. *ferulae* and *P. nebrodensis*.

Inzenga highlighted the excellent organoleptic qualities of *Agaricus nebrodensis* and reported the mushroom as very common on the crests of Madonie Mts, at 1900 m a.s.l., where I have collected in the feuds of Canna, Dragonara and in Pizzo dell'Antenna, on rotting debris of *Opopanax chironium* Koch and *Elaeoselinum asclepium* Bert., which grows in the valleys and at lower altitudes.

In the first case it is locally called Funcia di Basiliscu, and in the other case is named Funcia di Dabbisu.

Inzenga also reported: *Agaricus nebrodensis* is avidly collected by Communists in Polizzi Generosa, Bompietro, Castelbuono and Petralie, selling it at a price of two Italian Lire per kilogram, with care being required by all gourmets. In the town of Castelbuono you find *Agaricus nebrodensis* at the price indicated above and the shopkeeper Carmelo Ciolino makes substantial shipments by parcel post, by guaranteeing full satisfaction of his clients. Also Mr. Luigi Failla living in Castelbuono prepared in jars this delicious mushroom, which it is possible to enjoy all seasons, as I did for my personal use several times. Among all mushrooms I have known, *Agaricus nebrodensis* takes primacy and its edible quality and delicious taste are held in high esteem.

Inzenga [18], in the first Centuria, reproduces the figures already published in 1863 in the Giornale del Reale Istituto di Incoraggiamento (Fig. **6**): a medium sized basidioma cut vertically (Fig. **7**), a basidioma not yet fully developed and small immature basidiomata that usually do not arrive to full development (Fig. **8**). Therefore the figures mentioned above are included in the protologue, and iconography no. 1 was designated as lectotype [20]. References to the presence in Sicily of *Pleurotus* growing on Umbelliferous plants (*Apiaceae*), together with brief notes on their morphology and dialect names used by the Sicilian population, are included in the works of the pre-Linnean authors, Boccone [21] and Cupani [22].

Particularly in the book of Cupani (Fig. **9**) *Pleurotus* mushrooms were reported as follows: Fungi umbilicum exprimentes, simul albi C.B.P. Fungi plures simul albi, ad arborum radices, esculenti. I.B. vulgaté Funci di Ferra, di Chiuppu, di rusedda (qui Alcami Funci Ebrei dicuntur), di Dabbisu, Basiliscu, ò di Curmi à pedi, ò zucchi d'Arvuli, ò d'Oliva.

This is the confirmation that, since 1600, the *Pleurotus* mushrooms were known to botanists and ordinary citizens. In particular, in the Sicilian dialect, the term Funci means mushrooms and the names Funci di Ferra, Funci di Dabbisu, and Funci di Basiliscu correspond to *P. eryngii* var. *ferulae*, *P. eryngii* var. *elaeoselini*, and *P. nebrodensis* respectively. Knowledge about mushrooms of the genus

Pleurotus growing on *Apiaceae* plants remained for a long time a secret kept by a small number of people living on the Madonie (northern Sicily).

GIORNALE

DEL

REALE ISTITUTO D'INCORAGGIAMENTO

DI AGRICOLTURA, ARTI E MANIFATTURE

IN

SICILIA

TERZA SERIE — ANNO I.

PALERMO 1863
TIPOGRAFIA MICHELE AMENTA
via S. Basilio, 35.

Figure 6: The inside cover of the scientific journal containing the original description of *P. nebrodensis*.

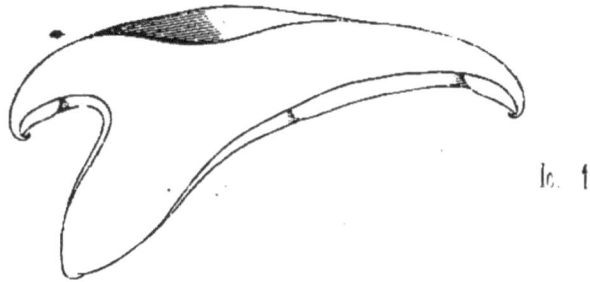

Figure 7: The original drawing (iconography no. 1) of the medium sized basidioma of *P.nebrodensis* included in the protologue and designated as lectotype.

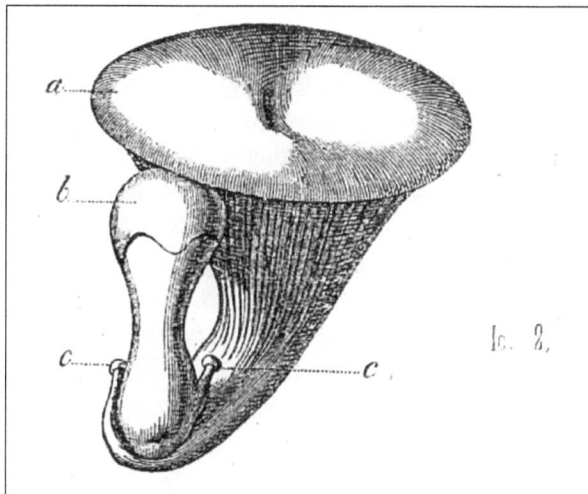

Figure 8: The original drawing (iconography no. 2) of a not yet fully developed basidioma with small immature basidiomata at the base.

Only in the mid-1800 Giuseppe Inzenga highlights in its publications morphological differences, and ecological distribution of *P. nebrodensis* and *P. eryngii* var. *ferulae*.

Many years later, Cailleux *et al.* [23] investigated inter-fertility and inter-compatibility relationships between different strains of *Pleurotus* collected in Sicily while Slezec [24] studied their karyology and highlighted a clear difference between *P. eryngii* var. *eryngii* and *P. nebrodensis*. The studies carried out by the above-mentioned French researchers do not take into account the large morphological and genetic variability of *Pleurotus* growing on *Apiaceae* plants, their ecological characteristics and the presence in Sicily of two other *Pleurotus* taxa; one with a white pileus (known by the farmers and mushroom pickers with the dialect name of Funci di Dabbisu) and the other with a brownish pileus (known by the farmers and mushroom pickers with the dialect name of Funci di Firrazzolu) respectively, only similar in appearance to *P. nebrodensis* and *P. eryngii* var. *eryngii*.

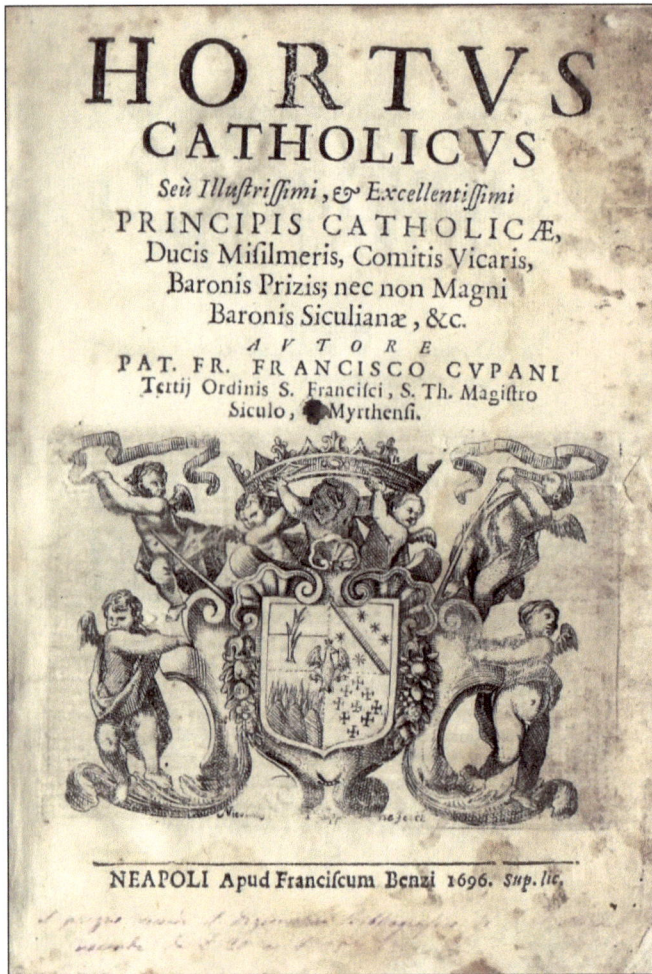

Figure 9: The inside cover of Horthus Catholicus, the iconographic work of F. Cupani.

For a correct interpretation of the taxonomy of *Pleurotus* growing on *Apiaceae* plants in Sicily, Inzenga's data must be carefully analyzed and critically evaluated as well.

The confusion which occurred over the years between *P. nebrodensis* and *P. eryngii* var. *elaeoselini*, comes from the fact that Inzenga [25] at the time of the original description of *P. nebrodensis* examined various mushrooms samples received as a gift from his friend, the Baron Turrisi Colonna, which were collected by farmers in different environments, from different associated plants and different altitudinal levels.

In fact, after a brief macroscopic analysis of the two fungi, the only noticeable difference is that the basidiomata of *P. nebrodensis* are heavier. The differences are most evident at the microscopic level as the basidiospores of *P. nebrodensis* are widely cylindrical and of larger size than those of *P. eryngii* var. *elaeoselini*.

In addition, *P. nebrodensis* possess sinuous cheilocystidia that are more or less knotty-lumpy and often bifid while in *P. eryngii* var. *elaeoselini* the cheilocystidia are attenuated at the apex and present, at times, a sharp beak. The two taxa can also be easily distinguished on the basis of their ecological preferences and distribution [26]. *P. nebrodensis* has a scattered distribution and grows at altitudes between 1200 and 2000 m only in spring, while *P. eryngii* var. *elaeoselini* has two fruiting periods throughout the year (March to May and October to November). After the publication of Inzenga's first Centuria *Agaricus nebrodensis* was reported as *Pleurotus eryngii* var. *nebrodensis*. In summary, the nomenclatural confusion was generated by Inzenga that in the original publication of *Agaricus nebrodensis* [25] reported the genera *Elaeoselinum* and *Opopanax* as potential plant hosts of *Agaricus nebrodensis*.

Besides Inzenga was also misled by his friend Baron Turrisi Colonna, who gave him the mushrooms received as homage by farmers.

Such white colored *Pleurotus* mushrooms were collected by farmers on the Madonie Mts at different altitudes and on different plant root residues (*i.e. Cachrys* and *Elaeoselinum*).

When Inzenga received the mushrooms by Turrisi Colonna he began to examine the basidiomata of both *P. nebrodensis* and of another white and macroscopically similar *Pleurotus* growing on *Elaeoselinum* root residues (now described with the binomial of *P. eryngii* var. *elaeoselini*).

Unconsciously Inzenga prepared mixed exsiccata of these two different taxa, and sent them to De Notaris, Fries and in other herbaria of European universities.

As reported by Venturella [20], a herbarium specimen from Inzenga, with a printed label, is kept in the Erbario Crittogamico Italiano (Fig. **10**), ser. II, in FI(!), and a duplicate of it is deposited in GE. Two other specimens, with handwritten labels by Inzenga are kept in GDOR.

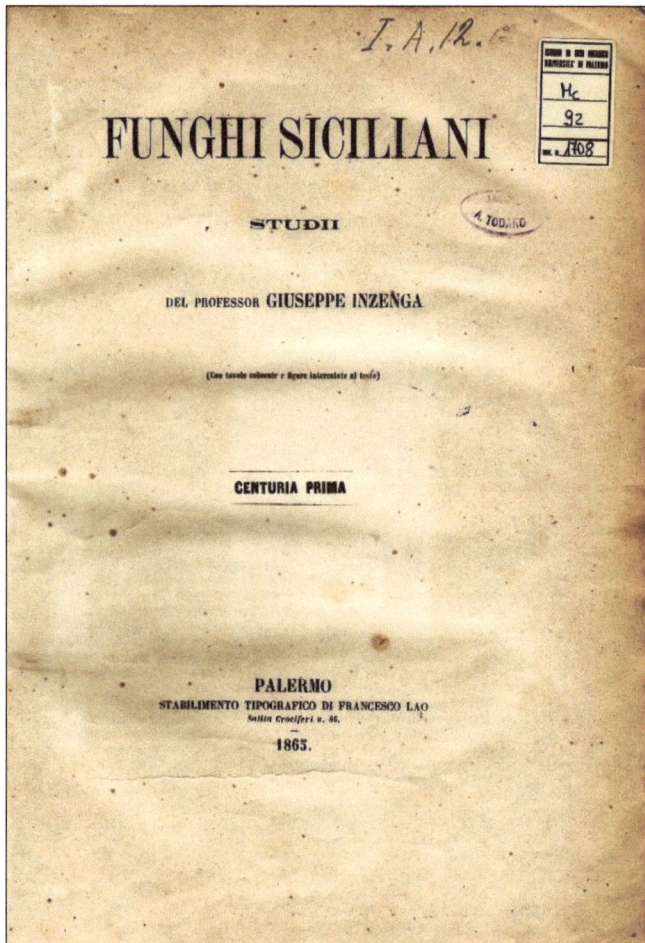

Figure 10: The inside cover of Inzenga's first Centuria.

A specimen kept in UPS(!) is undoubtedly part of the materials that Inzenga sent to E.M. Fries, but, as shown by correspondence between the two authors (Fig. **11**), subsequent to the original 1863 publication.

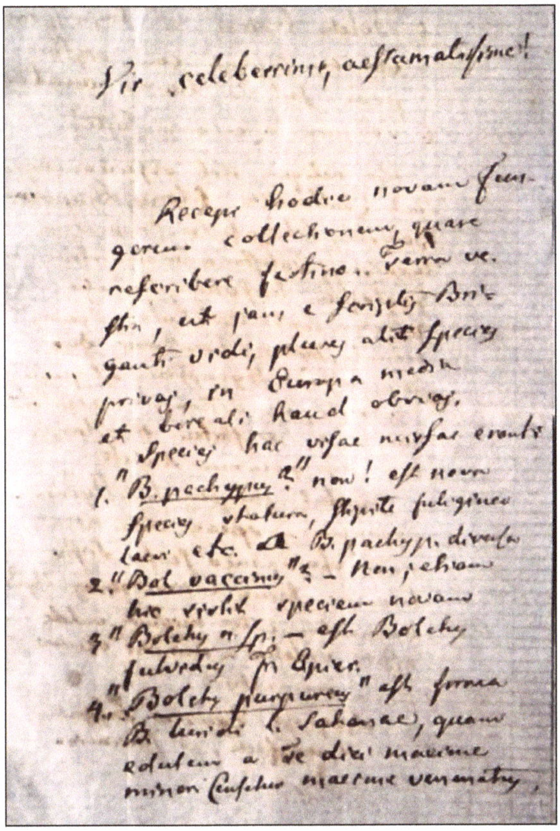

Figure 11: A letter of G. Inzenga to E. Fries kept in the Universitetbibliotek Handskriftsavdelningen of Uppsala (Sweden).

Three specimens, among which the centrally depicted one is quite similar to figure no. 1 of Inzenga's protologue, are present in PC(!).

The figure no. 1 of Inzenga's protologue was designated by Venturella [20] as lectotype, and the PC specimen placed in the centre of the herbarium sheet, as epitype (Fig. **12**).

The lectotype and epitype are both in good agreement with *P. nebrodensis* as currently understood, and also with Inzenga's original description, being both connected with Inzenga's name.

Through a careful interpretation of the comments of Inzenga, the drawings (Fig. **13**) included in the protologue and the descriptions reported in the Centurie, we

[27] were able to identify the historical collection sites of *P. nebrodensis* that Inzenga omitted from his publication, we provided an emended description of the species (Table **1**) based on numerous observation on fresh basidiomata (Fig. **14**), and we evaluated its declining fruiting productivity over the last centuries.

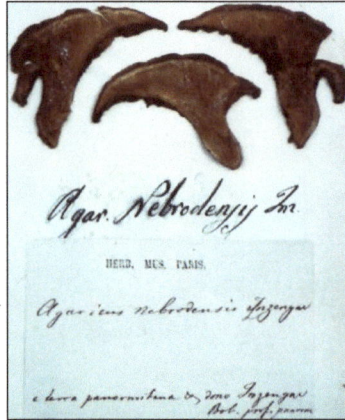

Figure 12: Herbarium specimens kept in the Muséum National d'Histoire Naturelle (PC, Paris, France), the central one designated as epitype.

Figure 13: The *P. nebrodensis* Inzenga's hand-made original color drawing.

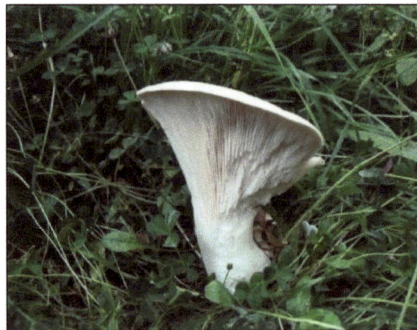

Figure 14: Basidioma of *P. nebrodensis*.

Table 1: Macro-morphological characters of *Pleurotus nebrodensis* based on a recent emended description [27]

Macro-morphological characters	
Basidioma	**Sturdy and Fleshy**
Pileus	3.0–14.5 cm (width), 3.7–13.5 cm (length); applanate, uplifted, shallowly depressed or convex in side view; ovoid or conchate in top view; unicolorous, light ivory, sometimes bicolorous, cream to ivory.
Margin of pileus	Plane, incurved, uplifted, incurved or involute in cross section, entire to erode in surface view. Surface of margin entire or eroded, smooth.
Surface of pileus	Shiny or translucent, dry, cracking-glabre, smooth or glabrous.
Flesh	Cream, with consistency hard-tough to turgid, without color changes, sulphur-yellow when dry, 1–2 mm thick at the margin and 1–4 cm thick at the center.
Taste	Mild and farinous.
Lamellae	4–8 mm width, 2.5–7.5 cm length, attachment adnexed to decurrent, gills spacing subdistant to close, moderately broad in thickness, light ivory colored, margin of gills smooth to eroded, face of gills waxed, lamellulae present, extending one-half to one-third the length of gills.
Stipe	1.4–3 cm width, 2.1–7.5 cm length, terete in cross section, equal to bulbose, slightly tapered to tapered at the base in longitudinal view. Consistency fibrous, flesh solid to stuff. Stipe eccentrically or lateral attached to pileus, inserted in the root residues of *Cachrys ferulacea*, basal tomentum and veil absent.
Stipe surface	Smooth, light ivory colored.
Growth habit	Solitary or connate.
Type of basidiomata attachment	Stipitate.

The Original Description [25]: *Ag. magnus caespitosus, albus, vel dilute sub-flavus, pileo carnoso nargine revoluto, lamellis confertis lineari-lanceolatis, liberis, decurrentibus in stipite sublaterali, vs. basim permixtis. Fungi umbilicum exprimentes, simul albi C.B.P. – Fungi plures simul, albi, ad arborum radices, esculenti J.B. – Cup. H. Cath. pag. 80. Pileus junior laevigatus, albus, subumbonatus, demum dilute flavus, irregulari, modo ex epidermide diffracia rimoso-tessulatus, gregarious, caespitosus, aliquando ob coacervata insitaque individua ramosus: 2–5 unc. latus, et ultra. – Stipes rare centralis, supra dilatatus atque in pileo diffuses, brevis, subnullus, basi attenuates. Lamellae confertae tenues, lineari-lanceolate, longo decurrentes sub striarum forma vs. stipitis basim productae. Lamellulae numerosae, breviores lanceolatae, longiores postice rotundatae. Caro fibrosa, subtenax, saporis gratissimi, ac odoris farina molitae,*

albida, sicca dilute-flava. Sporidia alba. Agarico Eryngii DC. characteribus variis consimilis, sed magnitudine, colore albido pilei sporidiorumque, stipite breviore, lamellis confertis, angustis, lineari lanceolatis omnino distinctus. In montium culminibus Siciliae, Nebrodibus magis obvia e radici bus marcescentibus Elaeoselini Asclepii Bert., Opopanacis Chironii Koch., *etc.* Aprili, Majo nive dilabente. Esculentus.

Micro-Morphological Characters: Spore print light ivory to cream. Spores 12.5–15.1(–18)   ×   5.2–6.1   µm,   cream,   heterotrophic,   asymmetrical, phaseoliphorm, smooth, hyaline, guttulate. Basidia with basidioles, 4-spored, 40–50 × 10–11.5(–14)µm, sterigmata 3–4.5 µm (Fig. **15**).

Cheilocystidia 50–60 × 6.2–7.5(–9) µm, leptocystidia type, clavate, apex mucronate to capitulate (Fig. **16**).

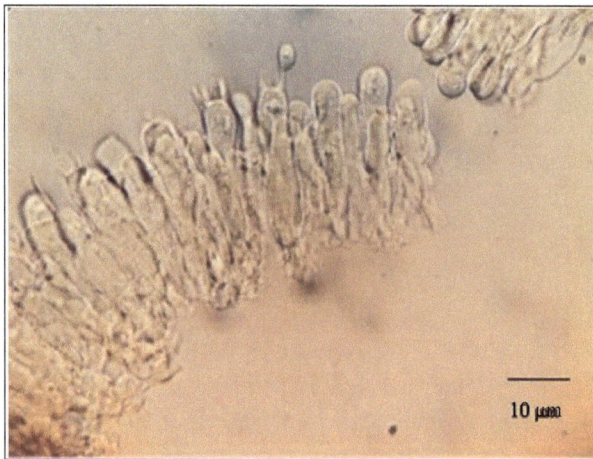

Figure 15: Microscopic view of basidia and spores of *P. nebrodensis* (resolution 60×).

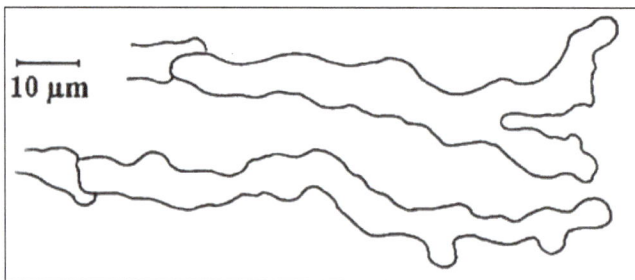

Figure 16: Cheilocystidia of *P. nebrodensis*.

Hyphal system monomitic. Hyphal wall thin. Hyphae septate with clamp connections (Fig. **17**).

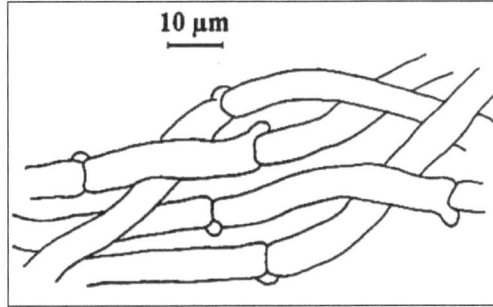

Figure 17: The hyphal system of hymenium in *P. nebrodensis*.

Specialized hyphae absent, no pigmentation. Pellis topography in two layers, 5–10 µm width, enterocutis, element of pellis absent (Fig. **18**).

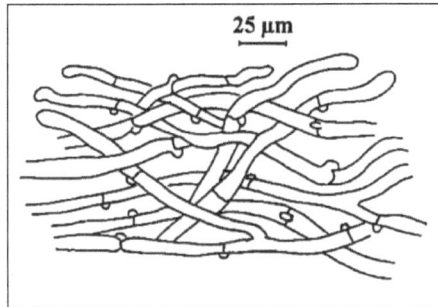

Figure 18: The pellis topography of *P. nebrodensis*.

The recent meeting with Mr. Norata (Fig. **19**), one of cattle farmers of Baron Turrisi Colonna, now a grey-headed gentleman in his 90s, enabled us to rediscover the historical collection sites of *P. nebrodensis*. Mr. Norata pointed out the sites which, at the beginning of the 20th century, were best for collecting this valuable mushroom. He also recounted his observation of the progressive decline of *P. nebrodensis*, which he attributed to intense exploitation of the area by cattle farmers and progressive loss of habitat due to road building.

On the basis of Mr. Norata's information and our field observations [27] we identified the two collection localities reported by Inzenga, which he named as "Canna" and "Dragonara".

Figure 19: Mr. Norata shows the historical collection sites of *P. nebrodensis*.

These are fiefdoms of the property of Baron Nicolò Turrisi Colonna's heirs. The two localities, nesting in a deep valley of the Madonie Mountains (northern Sicily) named Vallone Faguare, currently belong to the lands of Petralia Sottana (province of Palermo) (Fig. **20**).

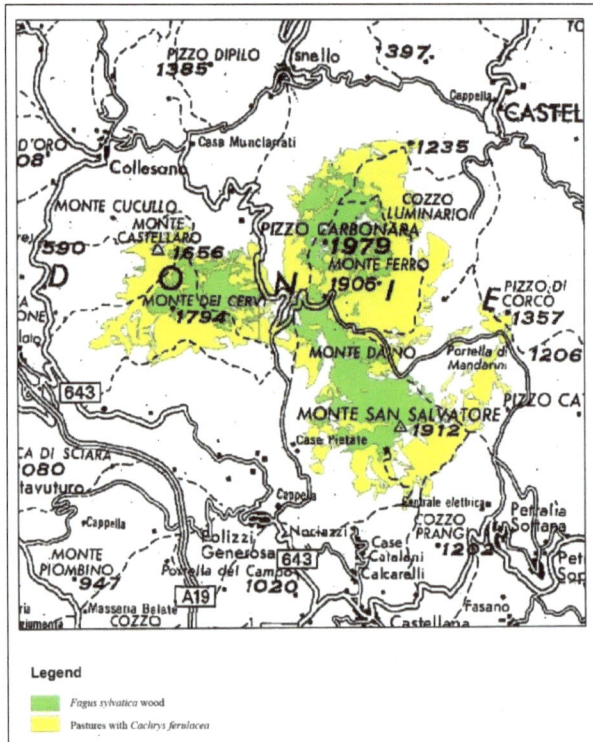

Figure 20: Cartography of the historical collection sites of *P. nebrodensis*.

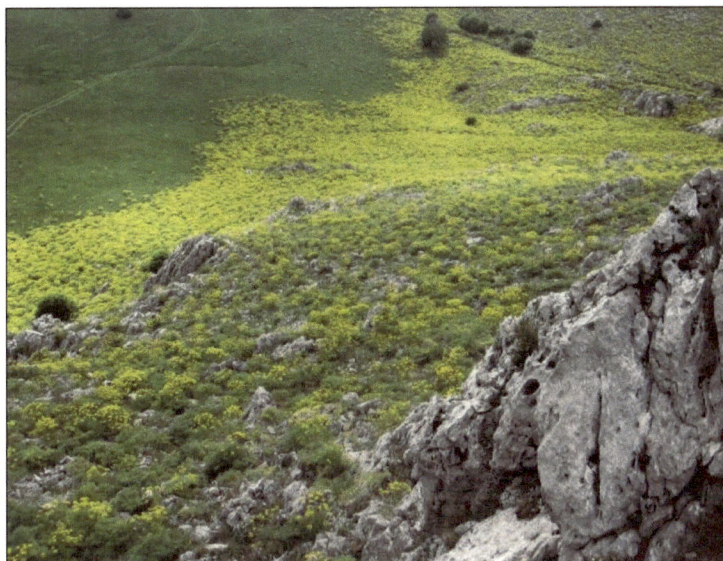

Figure 21: Flowering of *Cachrys ferulacea* in spring.

Their vegetation is characterized by wide-ranging pastures of *Cachrys ferulacea* (Fig. **21**) intensively exploited for grazing until 1989 when the Regional Park of Madonie was established.

ACKNOWLEDGEMENT

Declared none.

CONFLICT OF INTEREST

The author(s) confirm that this chapter content has no conflict of interest.

REFERENCES

[1]   Savastano L. Agrumi siciliani. Ann. Reale Staz. Sperim. Agrumic. Fruttic. Acireale 1915; 3: 5-10.
[2]   Muccioli A. Le strade di Palermo. Roma: Newton & Campton 1998; pp. 44.
[3]   Buttitta A. Relazione introduttiva. In Liotta G, Ed. I naturalisti e la cultura scientifica siciliana nell'800. Atti del Convegno; 1984 Dicembre 5-7; Palermo 1984; pp. 563.
[4]   Jackson BD. Guide to the literature of Botany. London: Longmans, Green & Co 1881; pp. 322.
[5]   Nordstedt CFO. Botaniska Notiser jör Ar. Lund 1889-90; pp. 29.

[6]   Saccardo PA. La Botanica in Italia. Materiali per la storia di questa scienza. Mem. Inst. Veneto Sci. Lett. Arti Venezia 1885; 25(4): 1-236.

[7]   Wittrock VB. Catalogus illustratus Iconothecae Botanicae Horti Bergiani Stockolmensis. Notulis Biographicis adjectis. Acta Horti Bergiani 1903; 3(2): pp. 198+46 pl.

[8]   Tucker EM. Catalogue of the library of the Arnold Arboretum of Harvard University. I. Serial publications. Authors and titles. Cambridge: Cosmos Press 1914; pp. 363.

[9]   Urban I. Geschrichte des könighlichen botanischen museums zu Berlin-Dahlem. Nebst auf zählung seiner samurlungen. Dresda: Verlag von C. Heinrich 1815-1913; pp. 360.

[10]   Rehder A. The Bradley bibliography. V. Index of authors and titles subject index. Cambridge: Riverside Press 1918; pp. 422.

[11]   Barnhart JH. Biographical Notes upon Botanist. CK Hall & Co, Boston: Massachusetts 1965, vol. II; pp. 549.

[12]   Stafleau FA, Cowan RS. Taxonomic literature. Le Bohn, Scheltema e Holkema, Vol. II: H, Utrecht 1979; pp. 991.

[13]   Venturella G. The contribution of Giuseppe Inzenga (1816-1887) to Sicilian mycological knowledge. In: Onofri S, Graniti A, Zucconi L, Eds. Italians in the History of Mycology. Accademia Nazionale delle Scienze detta dei XL and Società Botanica Italiana, Mycotaxon, Ltd 1999; pp. 89-96.

[14]   Briganti F. Historia fungo rum Regni Neapolitani, picturis ad naturam ductis illustrata, opus inchoatum a Vincentio Briganti, atque a Francisco Briganti, additis observationibus plurimsque figuris continuatum et in lucem editum. Napoli 1847.

[15]   Viviani D. I funghi d'Italia e principalmente le loro specie mangereccie, velenose p sospette, descritte e illustrate con tavole disegnate e colorate dal vero. Genova 1834.

[16]   Venturi A. I miceti dell'agro bresciano. Brescia 1860.

[17]   Vittadini C. Descrizione dei funghi mangerecci più comuni dell'Italia e dé velenosi che possono cò medesimi confondersi. Milano 1835.

[18]   Inzenga G. Funghi siciliani. Centuria I, II. Tipografia lao, Palermo 1965-9.

[19]   Inzenga G. Produzione svariata di funghi in Sicilia. Ann Agric Sicil 1877; n ser 96: 322- 4.

[20]   Venturella G. Typification of *Pleurotus nebrodensis*. Mycotaxon 2000; LXXV: 229-31.

[21]   Boccone S. Icones et descriptiones rariorum plantarum Siciliae, Melitae, Galliae et Italiae. London: Oxford 1674; pp. 96. 52 tables.

[22]   Cupani F. Hothus Catholicus, cum supplemento. Neapoli 1696. Idem supplementum alterum Panormi 1697.

[23]   Cailleux R, Diop A, Joly P. Relations d'interfertilité entre quelques représentents des Pleurotes des Ombellifères. Bull Soc Mycol France 1981; 97: 97-124.

[24]   Slezec AM. Variabilité du nombre chromosomique chez les pleurotes des ombellifères. Can J Bot 1984; 62: 2610-7.

[25]   Inzenga G. Nuova specie di agarico del Prof. Giuseppe Inzenga. Giorn Reale Ist Incorag Agric Arti Manifat Sicilia 1863; 1: 161-4.

[26]   Venturella G, Zervakis G, La Rocca S. *Pleurotus eryngii* var. *elaeoselini* var. nov. from Sicily. Mycotaxon 2000; 76: 419-27.

[27]   Gargano ML, Saitta A, Zervakis GI, Venturella G. Building the jigsaw puzzle of the critically endangered *Pleurotus nebrodensis*: historical collection sites and an emended description. Mycotaxon 2011; 115: 107-14.

Send Orders for Reprints to reprints@benthamscience.net

CHAPTER 6

Cultivation and Nutritional Value of *Pleurotus nebrodensis*

Maria Letizia Gargano[1], Georgios I. Zervakis[2] and Giuseppe Venturella[1,*]

[1]Department of Agricultural and Forest Sciences, Università de Palermo, viale delle Scienze 11, I-90128, Palermo, Italy and [2]Agricultural University of Athens, Laboratory of General and Agricultural Microbiology, Iera Odos 75, 11855 Athens, Greece

Abstract: A detailed report on *Pleurotus nebrodensis* cultivation methodologies, production costs, main pests and diseases and their control techniques is presented, together with information about the nutritional value and medicinal properties of the mushroom.

Keywords: Antitumor activity, *Agaricus bisporus*, *Apiaceae*, Control techniques, Cultivation methodologies, Diseases, Fatty substances, Food value, Genetic resources, Low-calorie diet, Medicinal properties, Mushrooms, Nutritional value, Oyster mushroom, Pests, *Pleurotus nebrodensis*, Production costs, Spawn preparation, Wild edible mushrooms, Vitamins.

INTRODUCTION

In Italy the most widely cultivated mushrooms are the champignons or "button" mushrooms [*Agaricus bisporus* (J.E. Lange) Imbach] (Fig. **1**) and the oyster mushrooms [*Pleurotus ostreatus* (Jacq.) P. Kumm.] (Fig. **2**).

In addition, and on a rather local scale, cultivation of *Agrocybe cylindracea* (DC.) Maire, *Pleurotus eryngii* var. *eryngii*, *P. cornucopiae* (Paulet) Rolland and *Stropharia rugosoannulata* Farl. ex Murrill is also performed.

The cultivation of "cardoncello" (*P. eryngii* var. *eryngii*) (Fig. **3**) started in the '70s; this agricultural activity could be implemented in open fields like those

---

*Address correspondence to Giuseppe Venturella: Department of Agricultural and Forest Sciences, Università of Palermo, vialedelle Scienze 11, I-90128, Palermo, Italy; Tel: +39 091 238 91 234; Cell: +39 329 615 60 64; E-mails: giuseppe.venturella@unipa.it; venturellagiuseppe1@gmail.com

existing in rural areas located in hills and mountains [1].

Figure 1: *Agaricus bisporus* in plastic containers before marketing.

Figure 2: Oyster mushroom cultivation.

Figure 3: The so-named "Cardoncello" mushroom (*P. eryngii* var. *eryngii*).

Furthermore, other mushroom species could be also cultivated by using the same technique, *e.g. P. eryngii* var. *ferulae, P. opuntiae, Polyporus tuberaster, Lyophyllum decastes, Ganoderma lucidum, Pholiota nameko* and *Agrocybe cylindracea.*

It is understood that a particular group of *Pleurotus* mushrooms (*i.e. P. nebrodensis*), known by the Sicilian population since a long time ago, presents organoleptic qualities superior to those of other wild mushrooms.

Recently undertaken cultivation tests are aimed at introducing *P. nebrodensis* in the Sicilian market by using production techniques different than those commonly adopted for the oyster mushroom.

The pursuit of this objective is also justified in the content of the conservation of biological resources and the knowledge that their protection can ensure the existence of a genetic pool, which could eventually provide food, medicines and other natural products together with additional benefits for rural communities.

The safeguarding of wild genetic resources of *Pleurotus* growing in association with Apiaceae plants is possible through the development of their commercial cultivation, which ensures a progressive decrease of human pressure on natural habitats.

Finally, the development of innovative methodologies in mushroom cultivation and the appearance of new products for immediate use may contribute at creating new financial (incl. employment) opportunities in economically depressed areas.

New innovative methods of *Pleurotus* mushrooms production were implemented for improving aspects of traditional cultivation techniques performed in specially designed chambers or modified greenhouses (Fig. **4**) as mushroom substrates are placed in light tunnels covered by a semi-shade nets, which are of low cost and easily constructed in the context of agricultural open fields [2] (Fig. **5**).

ISOLATION TECHNIQUES OF THE MYCELIUM AND SUBSTRATE PREPARATION

The cultivation of *P. nebrodensis* begins with the laboratory isolation of mycelium from mature basidiomata collected from natural habitats. The initial phase of development of the mycelium occurs by transferring, under sterile conditions, a piece of tissue taken from the inner part of the basidioma into a test tube or Petri dish containing the suitable culture medium (Malt Extract Agar, PDA, *etc.*) (Fig. **6**).

Figure 4: Cultivation of *Agaricus bisporus* in old railway tunnels.

Figure 5: Tunnel for mushroom cultivation.

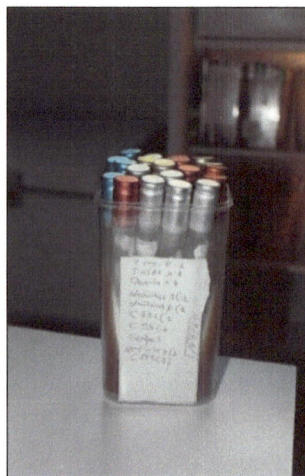

Figure 6: Test tube containing a culture medium for the development of *P. nebrodensis* mycelium.

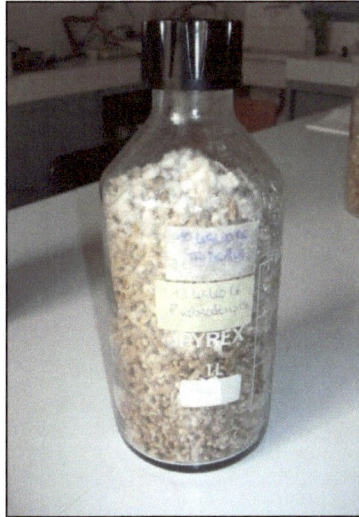

Figure 7: Sterilized wheat seeds (spawn) prepared in glass bottles.

After the development of the fungal colony, a small part of mycelium is used for the inoculation of sterilized wheat seeds in glass or plastic bottles with a capacity of one liter, where it can further develop and colonize the new substrate (Fig. **7**).

The material thus prepared ("spawn") is ready for inoculating the *Pleurotus* cultivation substrate.

The cultivation substrate is prepared with wheat straw and residues originating from the processing of sugar beet, moistened and mixed into a special machine (Fig. **8**).

Figure 8: Wheat straw and sugar beet residues moistened and mixed in a special machine.

Figure 9: Packaging of substrate in heat-resistant bags.

Then the substrate is transferred to a metering machine from which, manually or mechanically, it is packaged in heat-resistant bags weighing 4 kg (Fig. **9**).

The substrate bags are then placed in a basket and transferred into an autoclave where they are sterilized at a temperature of 110-120 °C for 1-2 h. After sterilization and cooling, the bags are transferred to a clean room where they are inoculated with spawn (Fig. **10**).

Figure 10: Clean room for mycelium incubation.

This process should be performed quickly and efficiently in order to avoid contaminations. After inoculation the bags are sealed and a sponge-like filter is placed on their top for permitting exchange of gases between the inner environment and the outside air.

Alternatively, the mycelium can be grown on toothpicks (Fig. **11**), which with a special syringe, could be partially inserted into the bag, from where new mycelium starts to develop and rapidly colonizes the entire substrate.

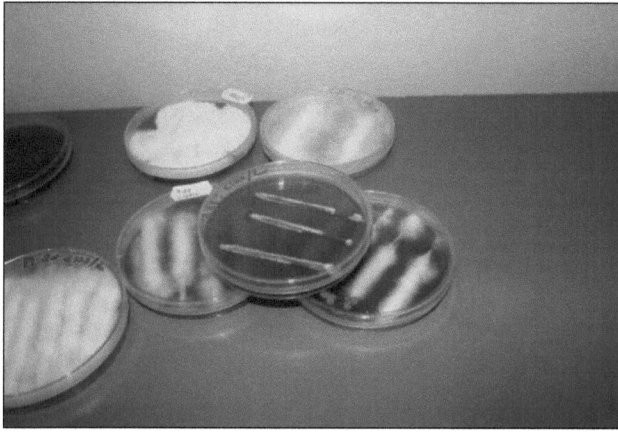

Figure 11: Cultivation of mycelium on toothpicks.

Inoculation may be also performed with liquid spawn, *i.e.* mycelium growing into a liquid medium.

The bags are then placed inside air-conditioned rooms, in the absence of light, where they are kept at a temperature of 25-28 °C; carbon dioxide concentration in the air does not exceed 1.5-2%. The duration of incubation ranges from 50 to 90 days depending on the technique used for inoculation, after which the substrate is ready to be used for the production of mushrooms (Fig. **12**).

CULTIVATION TECHNIQUE AND PRODUCTION COSTS

The cultivation of *P. nebrodensis* is realized into a tunnel-like construction, supported by metal arches, varying in length from 20 to 30 meters and covered with a black net providing ca. 90% shading [3].

In the inner part of the tunnel two or three cultivations beds are constructed, 1-1.10 meters wide, which are separated by aisles with a width of about 50 cm (Fig. **13**).

Figure 13: A cultivation bed of *P. nebrodensis*.

The bottom of cultivation beds could be prepared either slightly below the soil surface (by digging to a depth of 20-25 cm), or by forming an elongated heap on the ground level contained within wooden plates creating a peripheral wall around it with a height of 25 cm.

These structures could also obtain a more permanent character through the creation of concrete pavements onto which the cultivation beds could be formed.

The tunnel possesses also a humidification system with irrigation pipes and spraying nozzles arranged over the area of cultivation or a simple system for manual application of water (Fig. **14**).

When mycelium incubation is over and the substrate is fully colonized, bags are transferred into the tunnel where they are arranged, one beside the other, in the cultivation beds.

Then only the upper part of the plastic covering is removed and immediately after, the surface of the substrate is covered with a 1-2 cm layer of casing soil, taking care to fill up all empty spaces left between the bags (Fig. **13**).

The cultivation process itself does not require any complex operations but it is of fundamental importance to maintain the right levels of relative humidity, temperature and ventilation inside the tunnel in order to avoid any production losses that might be caused by pests and/or diseases.

During the first cultivation stage, before the formation of mushroom primordia, the cover soil must be always kept wet through the operation of the misting system (*i.e.* up to 4-5 mists of 5-10 sec. per day depending on the climatic conditions (Fig. **14**). If weather conditions are optimal, 8-10 days after the opening of bags and the coverage of the upper surface with the soil, mushrooms primordia start to appear (first flush). Then, after about a week, the second flush begins and the mushrooms will be ready for harvesting after ca. 10 days (Fig. **15**).

Figure 14: Distribution of water spray on substrate bags.

Figure 15: Early production of *P. nebrodensis* basidiomata.

The mushroom production generally takes place in two flushes, the first of which corresponds to about 70-80% of the total yield. Yields, under optimal conditions, are equal to approximately 28-30% of the weight of the substrate. The production per square meter of cultivated area is about 15-18 kg, while from a single bag of substrate, it is possible to obtain from 0.5 to 2 kg of mushrooms.

The duration of each production cycle is ca. 45 days after which, before proceeding to a new cycle, it is necessary to remove the spent substrates and to perform a disinfection of the room.

The purchase and installation costs for a standard cultivation tunnel with an area of 143 square meters (6.50 x 22 m), which is constructed with a galvanized steel body and it is covered by a plastic net (shading up to 90%), equal to € 1180.25 (€ 7.75 per square meter).

In most cases, each cultivation tunnel has two cultivation beds with a length of 22 meters each; thus a net total area of cultivation of 100 square meters is available for each tunnel. Given that each bag of substrate has a weight of 4 kg and that in each square meter of cultivation area 16 bags can be arranged, then the total quantity of substrate which can be placed within a tunnel is 6,400 kg. Since the purchase cost for each bag of inoculated substrate, including shipping costs, is € 3.10, for filling one tunnel (with 100 square meters of cultivation area which can 1,600 bags) a total cost of € 4960 is required.

Disinfection with a solution of formalin 2-3% (v/v) is required prior to cultivation for the materials used in the preparation of the cultivation beds (boards of wood or bricks). A preventive intervention can be performed after a few days after the planting of bags (usually with a product based on Prochloraz) by distributing 1-1.5 grams of active compound per square meter.

Mushroom picking is done by hand and the mushrooms are placed in wooden boxes, according to EU regulations, or plastic boxes with a capacity of 2.7 kg each and a unit cost of € 0.52 per box (Fig. **16**).

The mushroom harvest requires 8 hours (ca. 60-70 kg of mushrooms per hour are collected, cleaned and placed in a box), while the total number of working days varies from 19 to 22. Under optimal conditions, the average production of mushrooms varies from 1,152 kg to 1,536 kg per bag (18-24% of the substrate weight).

Figure 16: Mushrooms arranged in wooden boxes ready for sale.

Once the production is over, the spent substrate must be moved away from the cultivation tunnel. The removal and loading of the bags at the end of the cycle will take four working days at a total cost of € 180.76.

MAIN PESTS AND DISEASES OF CULTIVATED *PLEUROTUS* MUSHROOMS

Fungal Diseases

Hypomyces rosellus (Alb. & Schwein.) Tul. & C. Tul. is a pathogen producing a white to gray mycelium with a high speed of growth. In the early stages of development it is detected by the appearance of a thin mycelial mat. Later on, numerous conidial fruiting structures emerge from this mat that, macroscopically, have a white chalky colour. The introduction of this fungus into the cultivation tunnel occurs through the casing soil, where the pathogen grows as a saprotroph.

Spread of the disease is facilitated by unprofessional or inexperienced conduct of workers during cultivation, especially as regards hygiene measures and early intervention when the first outbreaks of infection are noticed.

The mold develops on the soil especially if excessively wet, and particularly at the points where the water stagnates. The disease infects mushrooms in their early stages of development and causes post-harvest rots.

Controlling *H. rosellus* has to follow two lines of action: a preventive one, and another through the application of fungicides.

A fundamental preventive action is carried out by implementing the appropriate cultivation actions.

First, care must be taken in order to avoid excess levels of air humidity in the tunnel, the ground becoming too wet, as well as cases of reduced or non-existent air circulation. This is possible by ensuring a constant aeration within the tunnel and a decrease of the humidity levels by carefully monitoring water applications. In addition, collection and removal of all mushroom residues from past cultivations should be performed since they represent an excellent substrate for the development of *H. rosellus*.

It should be remembered that *H. rosellus* is a pathogen that can be eradicated or even controlled with difficulty, while capable of affecting vast areas of cultivation in a short time. Therefore the defense against this disease has to be implemented mainly through preventive actions.

The most effective fungicide is Prochloraz (Sporgon), to be used as a preventive measure, in applications made 4-5 days after casing.

The term "green molds" (*Trichoderma* spp.) comprises diseases caused by Trichoderma viride Pers. var. viride and *Trichoderma koningii* Oudem.

The green molds can affect the substrate at the time of spawning or later, when the *Pleurotus* mycelium is already under development.

In the first case, the mycelium of the pathogen spreads rapidly across the substrate forming a grayish-white, moderately compact matt, which progressively obtains a powdery look that changes color from green to dark green as the fungus produces conidial fructifications.

The growth of the pathogen's mycelium causes a significant lowering of the pH to acidic values that are a further impediment to the growth of *Pleurotus* mycelium.

If the mold is established during the incubation period, then spread of the infection occurs in more or less extensive areas where the green conidial fructifications are mixed with the pure white mycelium of Pleurotus.

This condition may persist until the end of the incubation when the whole mass of substrate could undergo a limited retraction, assume a dark-green color, and become very compact with the presence of exudates at the bottom of the bag.

In other less severe cases, infected substrates could provide inferior yields in the first flush and no production in the second one, since by then the substrate is usually completely invaded by *Trichoderma*.

Other non aggressive *Trichoderma* strains can invade the substrate and limit their growth in corners at the base of the bag where there is a higher concentration of CO_2 and humidity.

These molds are overwhelmed by the *Pleurotus* mycelium growth and therefore they usually do not have a negative impact on production.

Again, for green molds, prevention is of significant importance for the control of the pathogen.

Bacterial Diseases

"Yellowing" is an infection caused by bacteria of the genus *Pseudomonas* (notably *P. tolaasii*) appearing on mushroom pilei and typically expressed with isolated, depressed, circular, yellow ochre-orange, dark hazel or dark orange-rust blotches with a diameter of a few millimeters.

The bacterial disease can also affect the stipe manifesting itself in the form of more or less elongated streaks, sometimes depressed, with shades of the same color to those appearing on the pileus.

In conditions of high humidity, pilei are covered with a thin layer of a slime-like liquid due to the development of bacterial colonies.

In section, mushrooms present deformed tissues in the stipe and pileus with gray-brown areas that tend to merge together creating extended deformations.

Mushrooms can be affected by bacteria in all stages of development. The bacterial disease may remain limited to a few mushrooms but it could spread quickly over large areas of cultivation. It could destroy one of the two production flushes without creating significant problems to the other or could cause significant yields decrease in both of them.

The factors affecting the onset of infection are mainly of environmental origin. The hot winds and high humidity are conditions that often determine the emergence of the disease to destructive levels. Other factors that favor the establishment of bacteria are excessive moisture levels and reduced or no exchange of air in the cultivation tunnel.

Although effective interventions are often difficult to be determined, it is recommended (as soon as the emergence of the bacterial disease is noticed) to suspend the administration of water and ventilate adequately the cultivation area. Mushrooms affected should be removed and applications with sodium hypochlorite solutions or a product based on iodine should be performed. In many cases, the severity of infection diminishes when climate conditions change. However, irrigation water should be checked for its microbial load and sanitary measures must be always taken together with proper moisture management.

Pests

Pleurotus mushrooms are easy prey for larvae of *Diptera* that cause infestations in both mycelium and basidiomata in various stages of development especially if the climate within the cultivation area is mild. The adults move on the ground or fly over mushrooms or casing soil. The females are attracted by the scent of the mycelium in the substrate, of the casing layer or of the basidiomata.

Insects penetrate into the stipe tissues and form tunnels as they feed. If the infestation develops during the early growth stages, then young basidiomata stop developing and become a spongy mass filled with larvae. When the mushrooms are mature, damage is limited to the central part of the pileus and at the point of contact between stipe and pileus. Larvae could also attack the mycelium, before the onset of mushrooms; in this case, a gradual depletion of hyphal biomass occurs in the substrate.

In addition to the direct damage caused by the larvae, adults of these insects also cause indirect damage as they carry from one place to another fungal spores, mites, bacteria, nematodes, *etc*. *Mycophila speyeri* Barnes is among the most active agents of infestation in the cultivation of *Pleurotus nebrodensis* in Sicily. Adult females and orange larvae are 1.5 to 2 mm long, and appear in just 7 days at temperatures of 25-28 °C.

The larvae quickly destroy the mycelium in the deeper layers of the bag. When substrates are irrigated, larvae tend to rise to the surface, spreading around and often forming groups along the stipe of the mushroom. The flies could breed on decaying plant debris, on wild mushrooms, on dung and on waste materials and proliferate at the optimum temperatures for mycelium growth (27 °C). The life cycle of these pests at 24 °C varies from 8 to 21 days (sciarid). These pests could cause the total destruction of *Pleurotus* production. As soon as the bags are placed into the cultivation tunnel, the scent of the mycelium is appealing to the adults for spawning. The larvae, as mentioned, hide deep in the substrate and initially only an abnormal delay in the appearance of mushrooms can be noticed. When irrigations are performed, the larvae are stimulated to move up onto the surface of the covering soil where they form a whitish or orange layer.

In many cases, flies infestations are attributed to the presence of substrate residues from previous crops in the cultivation tunnel.

Controlling infestations caused by *Diptera* is mostly based on the adoption of legislation related with preventive measures. Of considerable importance is the cleaning of the cultivation tunnel and of the surrounding areas as well as removing and destructing any residues and leftovers from previous cultivation.

Control with chemicals is allowed to a preventive level with distribution of the active compounds on the soil cover at the time of casing (Cyromazina, Methoprene, Deltamethrin).

After a severe attack of flies, it is recommended before starting a new crop cultivation to treat the surface with a solution of 2% formalin by distributing 0.5 to 1 liter per m^2.

The treated surface must be covered with a sheet of polyethylene for 24-48 hours.

Non Parasitic Abnormalities

Alterations in the morphology and in the normal appearance of *Pleurotus* basidiomata are usually due to abnormal conditions of humidity, temperature, ventilation and light, which occur during the cultivation phase.

Damage Caused by Excess of Moisture

Excess moisture in the casing layer makes it too compact and poorly ventilated. Under these conditions basidiomata may develop without a distinct pileus and stipe. Immediate reduction of the amount of water administered could attenuate the problem and the new mushrooms emerging are usually entirely normal as regards their morphological characteristics and organoleptical quality.

Damage Caused by Excessive Temperature

It is an alteration that occurs in spring and autumn mainly in mushroom cultivation units situated in south Italy. Abnormal differentiation of mushrooms appears in the form of irregular clusters, which are flattened, mixed with the soil cover, at first white then yellowish consisting of several indistinct basidiomata.

The damage is attenuated by lowering the temperature inside the tunnel through frequent mists on the outside part of the shading net.

In addition, of positive effects is (if implemented at the end of the warm season) increasing the thickness of the casing material on deformed mushrooms by placing a layer of soil of 1-2 cm, and taking care to keep it moist; hence, the growth of new mycelium is stimulated for producing new basidiomata.

Damage Caused by Reduced Ventilation

Climatic conditions with temperatures below 10 °C or the action of the wind may lead farmers to decide covering of the cultivation tunnel. In such cases, and especially when such a situation persists for a long time, there is an accumulation of carbon dioxide inside the tunnel that causes young basidiomata to assume a light ocher color and a flaccid consistency. In other cases, the mushrooms grow by forming swollen and elongated stipes and cylindrical small pilei that are not well differentiated. In mature mushrooms, the stipe becomes very elongated, while the pileus folds upwards assuming a funnel-like shape. This type of damage may be more frequent in crops where the growing conditions cannot be controlled with accuracy and/or where more experience is needed in adjusting the environmental parameters.

Malformations

This is a group of disorders, of an -often- unknown origin, which causes alterations in the morphology of cultivated mushrooms.

Appearance of Warts on Pilei

Warts occur on the surface or at the margin of the pileus in the form of small sessile or pediculate obtrusions, isolated or combined into groups.

The alteration is caused by sudden drops in temperature that occur during the early stages of cultivation when the mushrooms are rapidly growing.

The quality of the mushroom is not significantly changed, while the appearance is such as to prevent the marketing of the product.

Deformations with Spots on the Pilei

This is a deformation that occurs mainly when, after placing the bags into the cultivation tunnel, a sudden drop in temperature is accompanied with high rainfalls. The mushrooms grow in clusters and the pilei become brown-gray with dark spots.

This alteration can affect all mushrooms or just some, while others remain intact.

In some cases, when favorable climatic conditions are back, the growth of the clusters stops and the production returns to normal. The mushrooms initially seem little deformed, but later in the course of development, look almost normal, so that they can be marketed.

Mushrooms Deformations

The alteration is due to abiotic factors related to a particular susceptibility of the mushroom variety used.

For example, deformations are particularly evident in light-colored mushrooms, grown in cold climates or under high soil moisture conditions.

Mushrooms can take the form of a globular mass variously irregular which differs in thickness, while maintaining a light gray tint and a pruinose appearance.

They could also present stocky and enlarged stipes, pilei variously curled, and lamellae surfaces strongly reduced and deformed.

In other cases, mushrooms possess distinct stipes and pilei but their margins are curled and twisted.

Control Techniques

Effective control of pests and diseases in mushroom cultivation is usually very difficult, while irrational applications of pesticides could be dangerous to human health.

Most of the diseases are discernible only when they have assumed a significant dimension and an intensity that is no longer controllable, or many of them have a capacity to spread within a very short time putting at risk the entire cultivation process.

For this and other reasons, such as the need to minimize any possible contamination of edible products with chemicals, control of mushrooms pest and diseases is based primarily on prevention through careful and consistent application of all rules that allow maintainance of maximum hygiene conditions in the cultivation environment.

In particular, it is essential to carry out thorough and frequent sanitation of all areas of production and clean every day all equipment used.

Of the greatest importance is also the scholastic removal from the cultivation tunnel and its surrounding areas of all waste materials (old and discarded mushroom parts, substrate residues, *etc.*).

Finally, it is necessary that workers take careful personal hygiene measures associated with the use of clothing that is frequently renewed. If this preventive behavior becomes a routine in everyday life of a mushroom cultivation unit, then many infestation problems that threaten this particular crop can be greatly attenuated.

Despite the adoption of such preventive measures, the appearance of infestations cannot be excluded since intensive production and the particular environmental growing conditions could create adverse situations that need additional actions.

Therefore, in a few cases, it might be necessary to intervene with pesticides. The fact that mushrooms could absorb chemicals applied during their cultivation together with their short production cycle, are indicative of the need for the enactment of stringent/strict operational standards to prevent the presence of toxic residues for humans in the final product.

The chemicals that could be used in most of the *Pleurotus* cultivation were:

a) Prochloraz, marketed under the name of Sporgon, is mainly used for prevention purposes and it is very efficient against major pathogens that attack cultivated mushrooms such as *Lecanicillium fungicola* var. *fungicola* (Preuss) Zare & W. Gams (previously *Verticillium fungicola*), *Hypomyces perniciosus* (previously *Mycogone perniciosa*), *Hypomyces rosellus* (previously *Dactylium dendroides*), *Chrysonilia sitophila*, *Trichoderma* spp., *etc.* Prochloraz's active compound belongs to the family of Imidazoles and acts by inhibiting the biosynthesis of ergosterols (source: Chemical Bra, Verona, Italy). Administration is recommended 10 days before harvest;

b) Cyromazine, marketed under the name of Armor, is an insecticide active against larvae of *Diptera*. Administration is recommended until 14 days before harvest;

c) Deltamethrin or Decis Blue, is an insecticide active against larvae and adults of *Diptera*, whose treatment should be discontinued 3 days before harvest;

d) Metaldehyde is the active ingredient used for controlling snails, it can be found in the market in various formulations with a safety period of 20 days;

Disinfectants are used in the disinfection of rooms and equipment, they are active against bacteria, viruses and fungal pathogens. Those authorized in the cultivation of mushrooms are Environ (based on synthetic phenols), Jodothen-25 (iodine-based) and the PVP179 (also iodine-based, and used as bactericide, to be added to irrigation water, with preventive functions).

NUTRITIONAL AND MEDICINAL VALUE

The chemical composition of *Pleurotus nebrodensis* was analyzed on wild and cultivated fresh basidiomata [4,5]. These analyses showed a good correspondence with the organoleptic characteristics and composition of nutrients in the wild and the cultivated mushroom ensuring the maintenance of a good quality standard.

This affinity is also confirmed by the percentage composition of fatty substances (Table **1**).

Table 1: Percentage composition of the fatty substance in wild and cultivated basidiomata of *P. nebrodensis*

	P. nebrodensis	
Amount per 100g of Edible Part	**Wild**	**Cultivated**
Dodecanoic acid	0.20	0.18
Tetradecanoic acid	0.72	0.90
Hexadecanoic acid	14.60	12.66
cis-7-Hexadecenoic acid	1.23	1.55
Octadecenoic acid	2.23	2.40
cis-9-Octadecenoic acid	40.20	41.30
cis,cis-9,12-Octadecadienic acid	38.70	39.70
Cis,cis,cis-9,12,15-Octadecatrienoic acid	0.08	0.16
Eicosanoic acid	1.90	1.16

P. nebrodensis mushrooms can be used in any type of low-calorie diet, for their food value, for the content of vitamins and minerals. Of special interest was the high content of vitamin B_{12} and riboflavin found in *P. nebrodensis* mushrooms (Table **2**).

Table 2: Composition of wild and cultivated basidiomata of *P. nebrodensis*

Amount per 100g of Edible Part	P. nebrodensis	
	Wild	Cultivated
Water	89.6	93.8
Proetins	1.50	1.58
Lipids	0.25	0.39
Carbohydrates (including dietary fiber)	2.27	3.22
Ashes	0.77	1.01
Vitamin D_3 (µg)	0.20	0.26
Thiamine (mg)	0.018	0.027
Riboflavin (mg)	0.25	0.29
Niacin (mg)	4.3	5
Pantotenic acid (mg)	0.48	0.52
Pyridoxin (µg)	41	44
Biotin (µg)	18	18.3
Vitamin B_{12} (µg)	1.88	1.93
Ca (mg%)	50	52
Fe (mg%)	48	45
Mg (mg%)	130	92
K (mg%)	995	788
Na (mg%)	958	788
P_2O_5 (mg%)	915	588

Screening for the presence of metabolites particularly active against the cell metabolism that may have important therapeutic implications were recently carried out.

In particular the antitumor activity of water extract of *P. nebrodensis* against colon cancer cell was evaluated. Results demonstrate that extracts are able to modulate cell growth, apoptosis, homotypic and heterotypic cell-cell adhesions and signaling pathways [6].

ACKNOWLEDGEMENT

Declared none.

CONFLICT OF INTEREST

The author(s) confirm that this chapter content has no conflict of interest.

REFERENCES

[1]   Ferri F, Zjalic S, Reverberi M, Fabbri AA, Fanelli C. I funghi. Coltivazione e proprietà medicinali. Bologna: Edagricole 2007; pp. 271.

[2]   Zervakis G, Venturella G. Mushroom breeding and cultivation enhances *ex situ* conservation of Mediterranean *Pleurotus taxa*. In: Engels JMM, Ramanatha Rao V, Brown AHD, Jackson MT, Eds. Managing Plant Genetic Diversity. UK: CABI Publishing 2002; pp. 351-8.

[3]   Venturella G, Ferri F. Preliminary results of *ex situ* cultivation tests on *Pleurotus nebrodensis*. Quad Bot Ambientale Appl 1994-1996; 5: 61-5.

[4]   La Guardia M, Venturella G, Venturella F. On the chemical composition and nutritional value of *Pleurotus taxa* growing on umbelliferous plants (*Apiaceae*). J Agric Food Chem 2005; 53: 5997-6200.

[5]   Palazzolo E, Venturella G. Comparative analyses of the chemical composition of spontaneous and cultivated sporophores of *Pleurotus nebrodensis*. Quad. Bot. Ambientale Appl. 5(1994): 75-8.

[6]   Fontana S, Flugy A, Cannizzaro A, *et al.* Antitumor activity of water extract of *Pleurotus* species growing on root residues against colon cancer cells. Associazione Italiana di Biologia e Genetica, Assisi: Italy 2012.

Send Orders for Reprints to reprints@benthamscience.net

CHAPTER 7

Market Outlook, Production Chain and Technological Innovation for *Pleurotus nebrodensis* mushrooms

Maria Letizia Gargano[1], Georgios I. Zervakis[2], Alessandro Saitta[1] and Giuseppe Venturella[1,*]

[1]*Department of Agricultural and Forest Sciences, Università de Palermo, viale delle Scienze 11, I-90128, Palermo, Italy and* [2]*Agricultural University of Athens, Laboratory of General and Agricultural Microbiology, Iera Odos 75, 11855 Athens, Greece*

Abstract: This chapter provides a report of a marketing investigation performed in synergy with private companies in order to activate a production chain for the cultivation and trade of *Pleurotus nebrodensis*. A recent experience of technological innovation carried out in a farm of the Madonie (N. Sicily) territory is also presented. Notes on the status of Critically Endangered species and pertinent legislation are also provided.

Keywords: Conservation techniques, Consumers, Cultivation techniques, *Dactylium dendroides*, IUCN, Labeling, Legislation, Madonie Regional Park, Madonie Mts., Market, Monitoring system, Mushrooms, Packaging, *Pleurotus nebrodensis,* Production chain, Red List, Sicily, Technological innovation, Traditional recipes, *Trichoderma* spp.

INTRODUCTION

Since the origins of agriculture, dating back approximately ten thousand years ago, hundreds of plants have been cultivated which nowadays constitute an important food reserve for humanity. With modernization of agriculture, food diversity is concentrated within a selected number of species and varieties. This process has been greatly enhanced in the 20th century through the development and/or promotion of high-yielding varieties/hybrids in response to the

---

*Address correspondence to Giuseppe Venturella: Department of Agricultural and Forest Sciences, Università of Palermo, vialedelle Scienze 11, I-90128, Palermo, Italy; Tel: +39 091 238 91 234; Cell: +39 329 615 60 64; E-mails: giuseppe.venturella@unipa.it; venturellagiuseppe1@gmail.com

Maria Letizia Gargano, Georgios I. Zervakis and Giuseppe Venturella (Eds)

evergrowing needs of maximizing food production and facilitating its trade. As an immediate consequence of this attitude, a large amount of genetic resources vanished or became obsolete [1]. The genetic resources of edible mushrooms are relatively poorly known and under-exploited, thus this category forms part of a large group of products called "minor" or "underutilized". Several fungal species have been only marginally used and only close to their center of origin or their secondary centers of diversity. *Pleurotus* mushrooms are particularly known as active decomposers of numerous lignocellulosic residues deriving from agricultural, industrial and forestry activities and for the production of edible basidiomata of high organoleptic quality. The prospects for an increase of their cultivation globally are extremely favorable; today *Pleurotus* production amounts to ca. 700,000 tons/year, *i.e.* about a quarter of the respective total world figure [2,3].

The cultivation of *Pleurotus* mushrooms is widespread in the Mediterranean region, reaching about 10-20% of the total production of mushrooms in the same area [4]. The possibility of a further increase of the mushroom growing sector is linked to the diversification of production and the use of taxonomically well identified strains.

A significant increase in the production of *Pleurotus* mushrooms seems quite possible, since it could be linked to eco-friendly cultivation practices. Moreover, *Pleurotus* mushroom production requires relatively low energy levels and rather limited inputs of labor. Another important factor is the ability to reuse spent cultivation substrates as cattle feed [5] or as a fertilizer for plants and soil amendment [6].

An increase of the total area of mushroom cultivation, through appropriate actions, could contribute substantially to the growth of the agricultural economy, particularly in the so-called "marginal" areas, while also conferring to the improvement of the trade balance within the agrofood sector.

LOCAL MARKET OUTLOOK

The quality of *Pleurotus* mushrooms growing on *Apiaceae* root residues, is particularly significant within the agri-food system in Sicily. These products show deep connections with the components of the geographical area from where they originate, and their organoleptic quality can easily differentiate them from other types of mass agricultural production.

Successful production of mushrooms at a commercial scale has such prerequisites as the existence of pertinent knowledge together with applied research actions followed by careful planning and a substantial capital investment. Once production begins, problems related to extreme weather conditions, infestations from pests and diseases and/or volatile market prices are not uncommon and farmers have to be able to cope with such difficulties.

Marketing of *Pleurotus* mushrooms could achieve adequate levels of efficiency due to their environmental, cultural and gastronomic value. It is also well known that product's demand depends primarily on prices, consumer's income and their preferences system [7].

The consumer analysis thus assumes a focal importance in the process of formulation and implementation of business strategies, as it provides the reference coordinates for the definition of the proposed sale [8]. Therefore the cognitive objectives of market departure are the identification of factors that influence purchasing decisions and the distribution of the target clientele to help choosing one or more segments to which the sale offer should be addressed.

The development of market research, the improvement of methodologies and survey tools designed to provide a better understanding of consumption, and consumers to guide their choices are necessary conditions for the agri-food system because it can thus compete effectively at the European level.

In order to clarify the future of *Pleurotus* mushrooms production in the Sicilian market two main categories of consumers were identified: the restaurants and individual consumers.

In the case of the restaurants, we wanted to distinguish between those who use only cultivated mushrooms, those who use indifferently wild and cultivated mushrooms, and those who use only wild mushrooms. As regards to individual consumers, we took into account three types of people: the so-called amateurs mycologists with high experience in wild mushrooms collection and in their consumption as well, the non-amateurs mycologists, and the undecided or indifferent. The first group includes those consumers who can be defined as "experts" for knowing how to select the right product. For these consumers, mushrooms are considered as a food to buy preferably either from mushroom production units or directly from wild mushroom pickers, and at a regular (ca. one week) basis. This is a heterogeneous group of individuals with the ability to pay above the average, and who are willing to sacrifice some of their free time in the preparation and cooking of food, usually in a very traditional manner.

The second group includes those consumers who are not connoisseurs of mushroom products. These are buyers who have difficulty in recognizing the organoleptic properties of individual products, or judge on the basis of their appearance and/or understand the differences between the various qualities in the market.

For this reason, such consumers are prone to buy mushrooms at retail outlets (greengrocers or supermarkets) and rely on the advice of the seller. In any case, they do not appreciate significantly mushrooms as food, and they limit their consumption to a few species among the best known. Particularly relevant is that at the time of purchase their prime attention is given to the price. These are people mainly committed to work away from their house, with little time or desire to cook.

The third group belongs to the indecisive and/or indifferent consumers. As concerns restaurants, the 60.9% of restaurants uses either wild or cultivated mushrooms, the 30.4% prefers cultivated mushrooms while the remainder consumes wild mushrooms supplied by trustworthy mushroom pickers.

As regards the frequency of consumption of the product, it should be emphasized that, people who eat only wild mushrooms, do so mainly during the period that they are collected. Only when there is an over-supply of this product, the surplus quantities are deep-frozen.

Most of those who use both wild and cultivated mushrooms, prefer to buy them fresh, and only 21.4% of them consumes also the frozen product.

As regards the acquisition of wild mushrooms by restaurateurs, it should be noted that they are counting primarily on mushroom hunters of their choice. Instead, in the case of cultivated mushrooms, restaurateurs prefer to buy directly from the mushroom production units and to a lesser extent from the supermarkets. The choice depends primarily on the prices and the availability of the goods.

The majority of the population considers that Sicilian mushrooms is a food of easy and rapid preparation and uses them in the making of condiments, as a side-dish beside the main food, or grilled, or even as a main dish.

Among *Pleurotus* mushrooms, *P. nebrodensis* is certainly the best known and appreciated for its organoleptic qualities.

This is particularly evident in the territory of Madonie where these mushrooms are collected and consumed since many years ago, but also outside of this area, and in particular in Palermo, where picking of wild mushrooms is of interest to many consumers.

Pleurotus mushrooms are appreciated for the taste and texture of flesh and, especially in the case of *P. nebrodensis* the dried form is much sought after since its flavor is further enhanced.

As regards the perception of individual consumers to mushroom products, most of them are somehow worried about their safety, even if these are cultivated, and before they buy they usually ask the vendors for relative information.

This "suspicion" is further confirmed by the low frequency of mushroom consumption; as much as two-thirds of the consumers eat mushrooms only once every two-three months.

As regards the types of products sold, significant discrepancies occur between the so-called amateurs and not amateurs. The first prefer fresh mushrooms, while non-amateurs are more oriented towards the purchase of dried mushrooms.

Individual consumers, unlike restaurateurs, prefer to buy the product either at the supermarket or at the greengrocer's shop. In this case, the choice depends on the easier access to and the availability of the goods as well as on the prices charged.

In this way, they do not have any difficulties in recognizing the freshness, the date of harvest and the provenance of the product.

The majority of the consumers use the mushrooms mainly for the preparation of condiments and garnishes. In any case, the mushrooms are considered as a food to be consumed not only in special occasions.

PLEUROTUS NEBRODENSIS: TWO TRADITIONAL RECIPES

Appetizer (Fig. 1)

Ingredients for 6 persons: 600 g of mushrooms (*Pleurotus nebrodensis*), extra virgin olive oil, lemon juice, salt, pepper, parsley and arugula (rocket, rucola).

Marinate for half an hour the raw mushrooms, cut into thin slices with the olive oil, the juice of two lemons, parsley, salt and pepper to taste.

Figure 1: Appetizer of raw *P. nebrodensis* mushrooms with parmesan cheese, salad and ricotta pie topped with extra virgin olive oil.

Noodles (Fig. 2)

Ingredients for 6 persons: 600 g of mushrooms (*Pleurotus nebrodensis*), 600 g of noodles, 2 cloves of garlic, extra virgin olive oil, butter, salt, pepper, parsley, Parmesan cheese to taste.

Sauté garlic, extra virgin olive oil and a knob of butter. Add the mushrooms, salt and pepper and continue cooking, adding a little water paste. Blanch the noodles mix with the sauce and sprinkle with chopped fresh parsley at the time. Serve the noodles and add some Parmesan cheese.

Figure 2: Noodles with *P. nebrodensis* mushrooms.

DEVELOPMENT OF A PRODUCTION CHAIN FOR *PLEUROTUS NEBRODENSIS*

The experience gained from the University of Palermo in the production of *Pleurotus* mushrooms with valuable organoleptic qualities, together with the expertise of the companies Buontempo scarl (Consortium Company with Limited Liability), Le Due Sicilie srl (Limited Liability Company) and Cometa, made possible to provide a useful start-up for actions aiming at the development of new mushroom cultivation techniques, at the management of the fresh product and at the processing of the surplus quantities. Improvements in these domains stem from the need for a more rational handling of cultivated mushrooms, because the manufacturer is forced to process considerable amounts of fresh product within a short time with a high risk of quality deterioration.

Test of Satisfaction

The starting point of the activity was to conduct test ratings in a panel of consumers and restaurateurs for fresh and stored mushrooms and assess the product's perceived value by the end customer.

Of the 70 businesses selected, 30 were holiday farms, 30 were restaurants and 10 were taverns.

A significant percentage of taverns, restaurants and holidays farms (60%, 50% and 66% respectively) has expressed interest in the product and wanted to test samples.

In the interviews performed, restaurateurs had an excellent impression of the product, which was collected in the best condition, free from diseases and delivered within 48 hours.

The price sounds "right" when it is around 10 Euro/kg, but in exceptional cases (*i.e.* products of excellent quality) the prices could reach 12-13 Euro/Kg. Since *Pleurotus* nebrodensis is clearly perceived as a "rare" product, it could be purchased at even higher prices (25-30 Euro/Kg).

All respondents stressed the importance of price/quality ratio and continuity of supply as the market is currently served by small operators which are not able to provide it on a regular/steady basis.

Competition mainly derives from *Pleurotus* ostreatus, available at a price of 2-4 Euro/kg and from *Boletus* species (frozen product, when available) presumably imported from abroad (Eastern Europe) at a price of 10 Euro/kg.

The restaurateurs seemed willing to accept the packaging of the product, so arranged as to prevent damage, in 2-3 Kg polyethylene boxes. Traditional recipes that had been proposed in the Madonie tasting panel, were prepared in order to assess their acceptance by potential consumers during Trade Fairs like those of Gently Salt in Palermo and the Congress-Exhibition on the Export of Sicilian Agri-food products organized in Turin.

Development of Conservation Techniques

The harvesting procedures required the use of plastic containers for ca. 3 kg of product, sanitized with jets of steam and stored in storage areas with no risk of contamination.

The harvest was performed by operators who wore latex gloves to prevent accidental contamination. The shelf life for the processing was evaluated through repeated testing of product samples in 48 hours from harvest and up to a maximum temperature of 10 °C. Based on the evidence gathered during the investigation, it was decided to focus the project in the preparation of a gravy with mushrooms, of mushrooms canned in oil, a mushrooms patè and of a dried product. The gravy sauce ingredients were tomato, onion, extra virgin olive oil, basil, and *P. nebrodensis* mushrooms. The container chosen was 314 ml glass jars. Stabilization was initially conducted by boiling at 100 °C for 15 minutes, but it failed to produce satisfactory results. A test conducted as described by the guide ISTISAN 96/35 of the Ministry of Health showed a slight development of gas within the container, a pH of 3.90, while the bacterial load was limited but still not optimal. When the treatment process was modified by subjecting the products to sterilization (*i.e.* autoclave at 121 °C at 1.1 atmospheres for 15 minutes) then the production of a sauce with optimal characteristics was achieved.

Regarding the mushrooms preserved in oil, the ingredients used were *P. nebrodensis* mushrooms, extra virgin olive oil, pepper, white wine vinegar, marjoram (*Origanum majorana*), garlic and fennel seeds. The containers chosen were the 314 ml glass jars and the 2700 ml catering/retail bulk tins. Initially, the product was not heat-treated and gas developed inside the container. Stabilization was therefore conducted through heat treatment in a cell at a temperature varying between 80 and 110 °C for a period between 5 and 90 minutes. The best result as regards the containment of the bacterial load was achieved with short treatments at a high temperature. Pasteurization at 80 °C is also suitable.

For greater assurance of the stability of product a heat treatment, has been added to the empty jars.

The quality of the flesh of *Pleurotus* mushrooms also permitted to cut them into slices before drying them.

This treatment allows the mushrooms to obtain a particular aroma that, according to many consumers, makes them equal to the dried mushrooms of the *Boletus edulis* group.

The drying process of mushrooms previously sectioned into thin slices with a thickness of approximately 1 cm was tested in an oven at different temperatures and drying duration.

Optimal results were obtained at 35 °C maintained for different time periods depending on the initial moisture content of the basidiomata.

The dried product was packaged in envelopes under vacuum with good results in terms of stability and conservation of the organoleptic properties in both cases.

Packaging and Labeling

The Cometa company has further supported the design phase with a study on the packaging of the canned product. It has been proposed, for reasons of optimization of business supplies, the standardization of the glass jar for canned sauces with the use of an "America" type container of 314 ml. The consulting company has therefore developed the labeling of jars.

TECHNOLOGICAL INNOVATION BY THE FARM INITIATED BY "BUONTEMPO SCARL" AND "LE DUE SICILE SRL": A RECENT EXPERIENCE

The project led by the company Buontempo scarl - Le Due Sicile srl in collaboration with the Department of Environmental Biology and Biodiversity of the University of Palermo has aimed at the acquisition of the necessary knowledge and the development of innovative techniques for the implementation of mushroom cultivation in the territory of Polizzi Generosa (Fig. **3**), a small town included in the Madonie Regional Park (N. Sicily).

The decision by the company Buontempo scarl & Le Due Sicile srl to concentrate efforts in the field of applied mycology was taken after the Department of Agriculture and Forestry of the Sicilian Region included in the list of local products the wild and cultivated mushrooms (2001). Mushrooms constitute, in fact, a still little studied component of biodiversity and they are a potential lever for the development of innovative agricultural and niche products.

Figure 3: Panoramic view of the farms Buontempo and Le Due Sicilie.

The recent public interest in mushroom production, due to a more effective disclosure of their nutritional properties, opens up interesting areas of the market, in addition to those determined by the pharmaceutical industry and the use of fungi for the bioconversion of agricultural residues and by-products.

The territory of the Madonie is suitable for non-intensive forms of exploitation of natural resources. For this reason, it is important to develop and consolidate niche business proposals for generating income in the territory without this leading to the creation of "invasive production" infrastructure.

This project led to the development of mushroom production techniques for *Pleurotus nebrodensis*, which, as local products, could form a significant example case within the Sicilian agri-food system.

The selection of the suitable wild genetic resources for mushrooms with valuable organoleptic properties (Fig. **4**), the verification of the genetic integrity of the crops, the study of environmentally-friendly farming techniques (unlike those adopted by traditional cultivation performed in caves or in tunnels), and the development of effective post-harvest processing techniques that preserve the quality characteristics of the fresh product, were the main objectives of this project. They hence represent the ground of challenge for the promotion of a mushroom production line able to generate environmentally sustainable wealth in protected areas such as the Madonie.

Figure 4: Mycelium of *P. nebrodensis* in a Petri dish.

In addition, the project aimed at testing different substrates for the cultivation of *Pleurotus* mushrooms for avoiding the purchase and transportation of ready-made substrates (based on wheat straw and residues of sugar beet cultivation, and shipped in 4 kg polypropylene bags) from distant regions, *e.g.* Emilia-Romagna, Basilicata.

The company Buontempo scarl & Le Due Sicile srl experimented with the use of locally available agro-industrial and agricultural waste as possible components of the cultivation substrate in order to evaluate their suitability at a low cost.

Besides, they produced bags smaller in size than those supplied from farms of Emilia Romagna and Basilicata.

Research has addressed the combination of wheat straw with beans gin waste, which are both available in large quantities in the Madonie area.

A mixture of straw and bean residues (10%) appears appropriate to ensure a correct basis of nutrients thanks to the proper C/N ratio.

The farmers evaluated also the possibility of subjecting the substrate to:

- Autoclaving;

- Semi-sterilization at low pressure;

- Thermal treatment of dry material;

- Steam Pasteurization

- Pasteurization by immersion in hot water.

Experimental tests indicated that the method of pasteurization is more appropriate since it requires less expensive infrastructure, has lower energy costs and ensures the survival of beneficial microorganisms. Hence, pasteurization guarantees that the dangerous "biological vacuum" conditions would not occur, which otherwise could favor the onset of diseases (incl. green *Trichoderma* mold) during the incubation phase of the mycelium.

For preparing the substrate, straw and other ingredients are cut into pieces of 3-6 cm, mixed and then water is added so that the mixture reaches moisture levels of ca. 75%. When ready, the substrate is loaded into the pasteurization tunnel and then heated by a current of steam until it reaches a temperature of 60 °C where it remains for a period of 6 to 10 hours. The pasteurization process allows the establishment of a thermophilous microbiota which permits a fermentation of the substrate to take place both before and (mainly) after the peak temperature phase.

Following pasteurization, the substrate is inoculated with spawn and polypropylene bags are filled under semi-sterile conditions. The bags are then closed with a synthetic sponge which allows exchanges of gases between the inside and outside environments (Fig. **5**).

For improving the cultivation techniques, *Pleurotus* mushrooms production was performed in shaded and not air-conditioned tunnel-like structures. The materials chosen consisted of a) "Dalmine" type metal tubes joined to the supporting structure; b) plastic mesh providing shading of 90%; c) sheets of polyethylene for the cover, and, d) wire mesh protection for avoiding possible entry of animals.

The bags of substrate were placed one beside the other inside wooden crates whose bottom was covered with a sheet of polyethylene.

Figure 5: Bags of compost closed with a synthetic sponge after inoculation.

The cultivation tests allowed to verify the correlation between climatic conditions and productivity with a clear and progressive slow-down of mushroom production when temperature decreases in autumn and when it increases at the end of spring.

During the cultivation tests performed in the tunnel, the substrate bags, and consequently mushrooms, suffered from problems caused by both biotic and abiotic factors. In particular, fungal diseases caused by *Dactylium dendroides* (Fig. **6**) and by green *Trichoderma* spp. molds were evidenced on the substrates, while yellowing of basidiomata was verified by pathogenic bacteria of the genus *Pseudomonas* as well as occasional pest infestations (in particular *Diptera* flies).

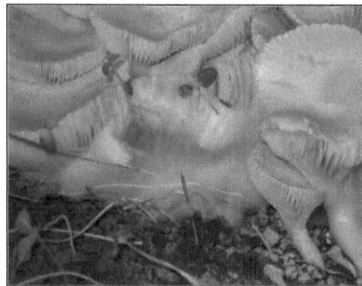

Figure 6: *Dactylium dendroides* in the pre-harvest stage.

Among the causes of product deterioration due to abiotic factors were sudden drops in temperature, wind action and, above all, over-accumulation of carbon dioxide (due to reduced ventilation). The latter was responsible for malformations of basidiomata presenting swollen and elongated stems and small under-developed pilei folding upwards assuming the shape of a funnel.

As regards fungal diseases a protocol was developed for disinfecting production environment, serving as an alternative to the traditional use of formaldehyde, which involves the use of injectors for the first cleaning of the crates. Subsequently a disinfection of the production area is carried out with 2% hydrogen peroxide at 10 volumes.

The colonization of basidiomata by larvae of *Diptera* was met with some success through the massive use of sticky traps that allowed to control the population of adult individuals within the growing environments.

The construction of stacked pallets permitted the doubling of the production area inside the tunnel.

Further improvement activities were conducted by adding casing on the top of open substrate bags. Initially plain sieved soil was used, but it was later replaced with semi-sterile composted soil in order to reduce exposure to pathogens.

To ensure the correct amount of air humidity inside the cultivation environment and into the mass of substrate, a fogging system controlled by a programmable electronic timer was installed.

Despite the fact that improvements have increased the performance of the tunnel without the use of air conditioning, it became clear that the environment in the tunnel could not be adequately controlled in this way without considerably risking productivity.

It was possible to verify that all strains suffered from unfavorable seasonal weather conditions as expressed by substantial halts in production when temperatures decreased or by deterioration of the product prior to harvest due to sudden increases in temperature caused by warm S-S.E. "scirocco" winds.

The project activities were therefore directed to the prototyping of a contained cultivation environment, air conditioned but with low energy costs, capable of supporting production policies oriented towards the generation of large quantities of high quality product. Therefore a cultivation unit was assembled able to exploit the availability of surface water for cooling and the capacity of solar panels to heat storage for warming up the unit.

The diagram shown in Fig. **7** illustrates the structure of the production unit.

The structure was assembled by the staff of Buontempo company with the support of external consultants.

The following figures show some of the stages of construction and the insulation of the coils for air conditioning.

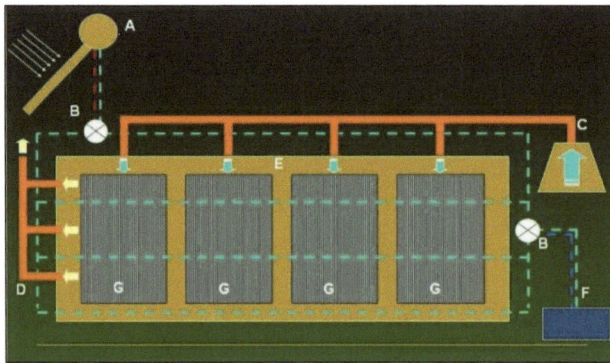

Figure 7: Construction scheme of the cultivation area. a) solar panel with battery; b) solenoid; c) adiabatic steam generator; d) extraction system; e) galvanized steel structure insulated with polyurethane foam and covered with polypropylene honeycomb panels with removable polycarbonate; f) accumulator of groundwater; g) removable panels made of Makrolon®.

Initially the cultivation unit was controlled in a simple manner through the placement of an integrated hygrometer to a relay for putting into operation the humidifier when RH exceeded a threshold RH level. The suction fan of CO_2 was driven by an electromechanical timer at intervals of 15 minutes. The temperature was maintained at about 20 °C by circulation of chilled water through the coils. However, growth of mushrooms was often accompanied by abnormalities in basidiomata formation, which were attributed to high concentrations of carbon dioxide.

The project was, therefore, commissioned to the firm Wisenet Engineering srl, for coming up with a system of continuous monitoring of environmental parameters during the critical phase of mushroom growth.

The availability of this information would, in fact, contribute at controlling with precision the environmental conditions during the production cycle. The network sensors would, in perspective, be able to control each parameter, and by means of dedicated hardware and software systems, to modify it according to the cultivation requirements. It was set to operate at predetermined time intervals so that the monitoring system could also optimize energy consumption.

As it is known from pertinent literature, the critical parameters for the development of *Pleurotus* basidiomata are:

- Air relative humidity (RH in the range of 80-90%);

- CO_2 concentration (500 to 1000 ppm);

- Temperature (around 23 °C);

- Illumination (at least 8 hours of artificial light).

For the measurement of the aforementioned parameters, the monitoring system (Fig. **8**) possesses sensors suitably placed in 5 different positions: three sensor nodes "RH/temperature/light" and two sensor nodes "CO_2" are located respectively in the three cultivation beds and in two distant positions within the cultivation area.

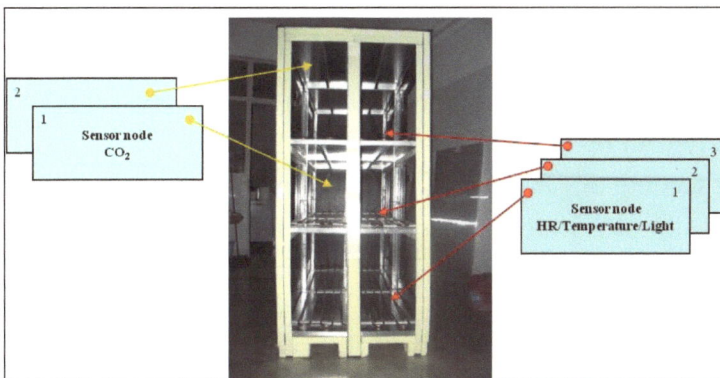

Figure 8: Distribution of monitoring sensors in the cultivation area.

Each of these sensor nodes acquires the respective physical quantities to be detected at regular time intervals; each acquisition is transmitted to the gateway node along with other information as reported in Table **1**.

Table 1: Information transmitted by the nodes of the monitoring system

NODE	Transmitted signal						
	ID	Tm	Bl	T	L	HR	CO_2
HR/T/L	X	X	X	X	X	X	
CO_2	X	X	X				X

Legend: HR = relative humidity; T = temperature; L = light; ID = device identifier; Tm = time measurement; Bl = battery level.

The ID code (device identifier) associates the number of the sensor node to the information provided (Fig. **9**).

An internal timer in each node turns it on at a certain time point, and as soon as it acquires the necessary information (which is in turn provided by the relevant transducers) these values are compared with the target values initially set by a microcontroller. When the measurements fall within the desired values-range, then the sensor node automatically shuts off.

In the case in which the measured parameters are out of the range initially set, then the sensor node remains lit and detects every 30 seconds (time set by software) the new information provided by the transducers. The mismatch between the detected values and target values ensures that the gateway node shares the relay for turning on one or both motors (fan and/or humidity generator). These motors remain on until all sensors confirm that the variables are within the target values, sending a "OK" signal to the gateway node, which then shuts them off. In addition, the sensor nodes ceases to detect the parameters sizes and remains off for the next 30 minutes.

Each measured value is associated with time/date, and stored in a data logger (together with all transmitted data), which can be accessed from the outside *via* a

USB port and hence accumulated data can be downloaded periodically on a PC or a handheld device.

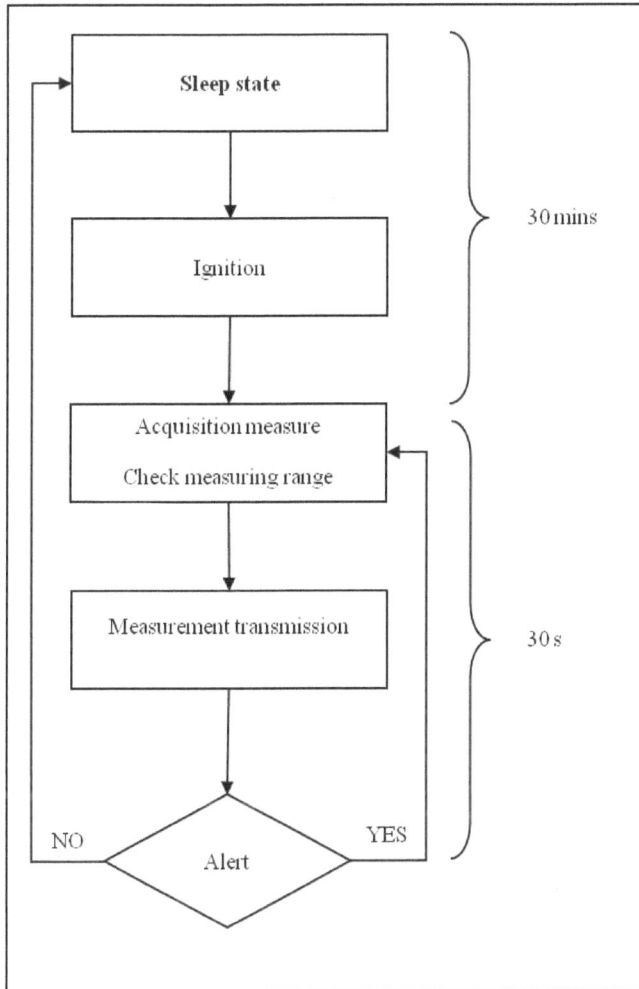

Figure 9: Flow chart of the measuring system.

Inside the cultivation unit (Fig. **10**), the following machinery is installed for monitoring and controlling the environmental parameters:

- Mushroom beds are equipped with fans which operate whenever the value of CO_2 exceeds a certain predetermined value which is monitored continuously by the two nodes dedicated sensor, positioned inside the mushroom bed. Each sensor communicates the data to the microcontroller at the gateway which will operate in a wireless mode the fan relay for turning it on or off so that the value in the environment does not exceed the threshold set for CO_2 concentration;

- A mist generator together with the relevant sensors (as described above for CO_2) permit regulation of RH values in a range between 80% and 90%. Even this humidifier system possesses a receiver that drives a relay for switching on and off the machine.

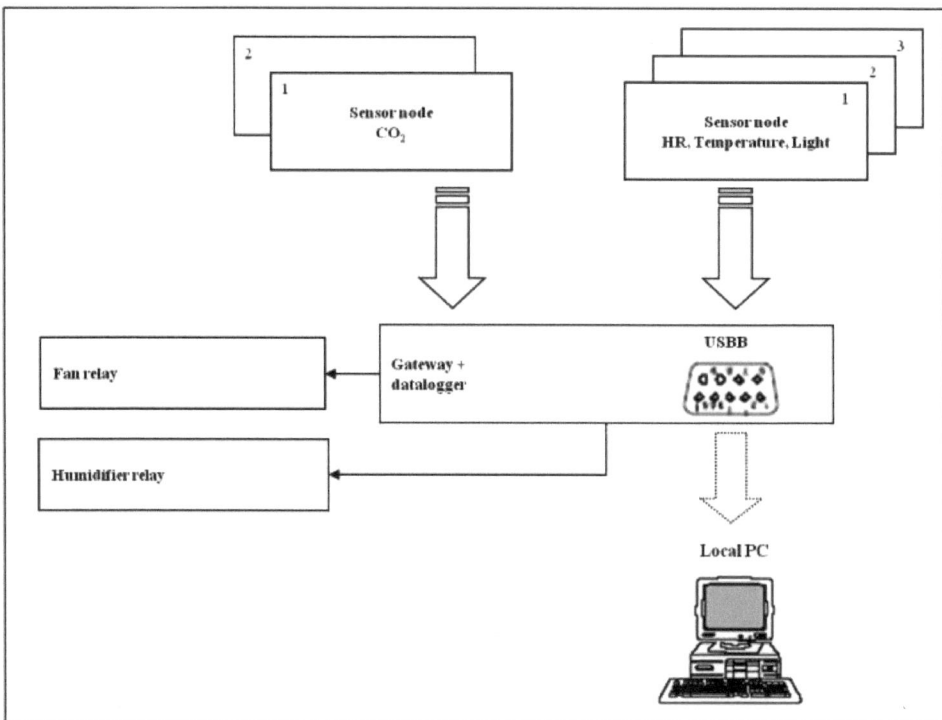

Figure 10: General diagram of the monitoring system.

CONCLUSIONS

Pleurotus nebrodensis is the only mushroom included in the IUCN Red List of Threatened Species and the "Top 50 Mediterranean Island Plants" [9].

This mushroom has been categorized CR (Critically Endangered) according to

IUCN Red List Criteria B1ab (iv, v)+2ab (iv, v).

In Sicily the collection and marketing of *P. nebrodensis* is regulated by the Regional Law of February 1st, 2006, no. 3 and by the subsequent Decree of the President of the Sicilian Region, August 4th, 2009. The Article 2 of the Decree prohibits the collection of basidiomata of *P. nebrodensis* with pileus dimensions less than 3 cm.

The Park of the Madonie has issued a set of regulations for the collection and sale of wild epigeous and hypogeous mushrooms. As reported in the paragraph 5 of Article 5 "Limitations for mushroom collection": "the collection of wild basidiomata of *P. nebrodensis* is prohibited in the Zone A of the Madonie Park, while in the other areas the collection of basidiomata of the same species is prohibited when pilei are smaller than 3 cm".

The paragraph 4 of Article 20 "Supervision and sanctions" imposes a sanction feevarying from 77.47 € to 154.94 € for mushroom pickers who collect basidiomata of *P. nebrodensis* smaller than 3 cm (Fig. **11**). The Article 14 allows the marketing of wild and cultivated basidiomata of *P. nebrodensis.*

Figure 11: Forest Service agents place a fine on mushroom pickers without pertinent authorization and they confiscate the mushrooms collected.

The recent actions of a legislative nature in support of agrotourism activities and those relating to the strengthening of the agro-food sector as well as the new regulations for the protection of areas with agricultural products of particular quality and unique characteristics have stimulated the interest of the farmers to become involved in their production.

These are defined as typical local products for their quality and nutritional properties, and have been over the years kept alive within rural communities thanks to the relentless transfer of traditions from generation to generation.

The demand and consumption of traditional local products is gradually increasing.

Nowadays, requests for traditional products exceed supply and for this reason the Italian Regional Administration undertook initiatives to accelerate the development of local production.

Wild edible mushrooms are one of the lesser-known elements of terrestrial biodiversity but, if properly studied and exploited, they can be used to increase and diversify the current agricultural/forestry production.

Attention of public opinion to the exploitation potential and the possible uses of wild edible mushrooms have gradually grown through a more effective dissemination, both by the scientific community and the public media.

The investigations carried out in the last ten years have shown that the cultivation of *Pleurotus* mushrooms in Sicily could play a very important role in rural development by generating (among others) additional income for marginal communities, at increasing mushroom consumption and at protecting wild populations of *P. nebrodensis*.

In conclusion, considering that the consumption of mushrooms is increasing, as well as the demand for food of high quality, the introduction of new items in the market, like *Pleurotus nebrodensis* mushrooms, that meet such requirements may be viewed favorably by the consumer with positive effects on social and economic development.

ACKNOWLEDGEMENTS

This e-book was carried out in the frame of a research grant awarded by the Italian Ministry of University and Research (MIUR) entitled "Caratterizzazione biomorfologica, ecologica, produttiva e qualitativa di *Pleurotus* nebrodensis (Inzenga) Quél., raro basidiomicete a rischio di estinzione/Bio-morphological, ecological, productive and qualitative characterization of *Pleurotus* nebrodensis (Inzenga) Quél., a rare basidiomycete at risk of extinction". Thanks are also due to Dr. Antonio Morello, a food technologist for the contribution provided for the development of preservation techniques, to Dr. Fabio Montagnino and Giovanni Faletra for making available facilities and technologies in their farm. We also wish to thank Ing. Mario Castelluccio for having kindly granted permission to take pictures of his company Italmiko.

CONFLICT OF INTEREST

The author(s) confirm that this chapter content has no conflict of interest.

REFERENCES

[1]   Gari JA. Conservation, use and control of biodiversity. Local regimes of biodiversity *vs.* the global expansion of intellectual property rights. Perspect Intellect Property 2001; 9. Special issue on: IP in Biodiversity and Agriculture.

[2]   Chang ST. Mushroom research and development-equality and mutual benefit. In: Royse DJ, Ed. Mushroom biology and mushroom products. Pennsylvania State University: University Park 1996; pp. 1-10.

[3]   Philippoussis A, Zervakis G. Cultivation of edible mushrooms in Greece: presentation of the current status and analysis of future trends. In: Van Griensven L, Ed. Science and cultivation of edible fungi. Baalkema: The Netherlands 2000; pp. 843-8.

[4]   Zervakis G, Venturella G. Mushroom breeding and cultivation favors *ex situ* conservation of Mediterranean *Pleurotus taxa*. In: Engels JMM, Ramanantha Rao V, Brown AHD, Jackson MT, Eds. Managing Plant Genetic Diversity. IPGRI, CABI Publishing 2002; pp. 351-8.

[5]   Tripothi JP, Yadar JS. Optimization of solid substrate fermentation of wheat straw into animal feed by *Pleurotus ostreatus* - a pilot effort. Anim Feed Sci Tech 1992; 37: 59-72.

[6]   Levanon D, Danai O. Chemical, physical and microbiological considerations in recycling spent mushroom substrate. Compost Sci Utiliz 1995; 3: 72-9.

[7]   Magni C, Mattiacci A. Gli stili alimentary europei. I risultati di una ricerca di mercato ed alcune riflessioni incomplete. Agribusiness Manage Amb 1995-96; 1(4).

[8]   Fiocca R. Rileggare l'impresa. Relazioni, risorse e reti: un nuovo modello di management. RCS Libri SpA: ETAS 2007; pp. 35.

[9]   Venturella G. *Pleurotus nebrodensis*. In: de Montmollin B, Strahm W, Eds. The Top50 Mediterranean Islands Plants. Wild plants at the brink of extinction, and what is needed to save them. IUCN Gland Switzerland and Cambridge: UK 2005; p. 98.

Index

A

Apiaceae 31-56, 57-62, 63-74, 75-98, 99-120, 121-143

Ascomycota 71, 3-30

B

Bacterial diseases 111

Basidiomycota 31, 57, 63, 71, 75, 3-30

Biodiversity 63, 130, 142, 3-30

C

Cachrys ferulacea 41, 58, 60, 61, 65, 66, 79, 93, 97

Convention on Biological Diversity 3-30

Cultivation 78, 31-56, 99-120, 121-143

E

Elaeoselinum asclepium subsp. *asclepium* 42, 58, 60, 84

Eryngium campestre 58, 59

F

Ferula communis 42, 43, 58, 59, 63

Fungal conservation 3-30

Fungal diseases 109

I

Inzenga 75-98

IUCN 3, 17, 18, 23, 140

M

Market outlook 121-143

Medicinal value 118

www.ingramcontent.com/pod-product-compliance
Lightning Source LLC
Chambersburg PA
CBHW041712210326
41598CB00007B/624